Pupil Book 5

Written by Jayne Campling, Andrew Jeffrey,
Adella Osborne and Dr Tony Wing

OXFORD

Contents

How to use this book

Welcome to Numicon Pupil Book 5.

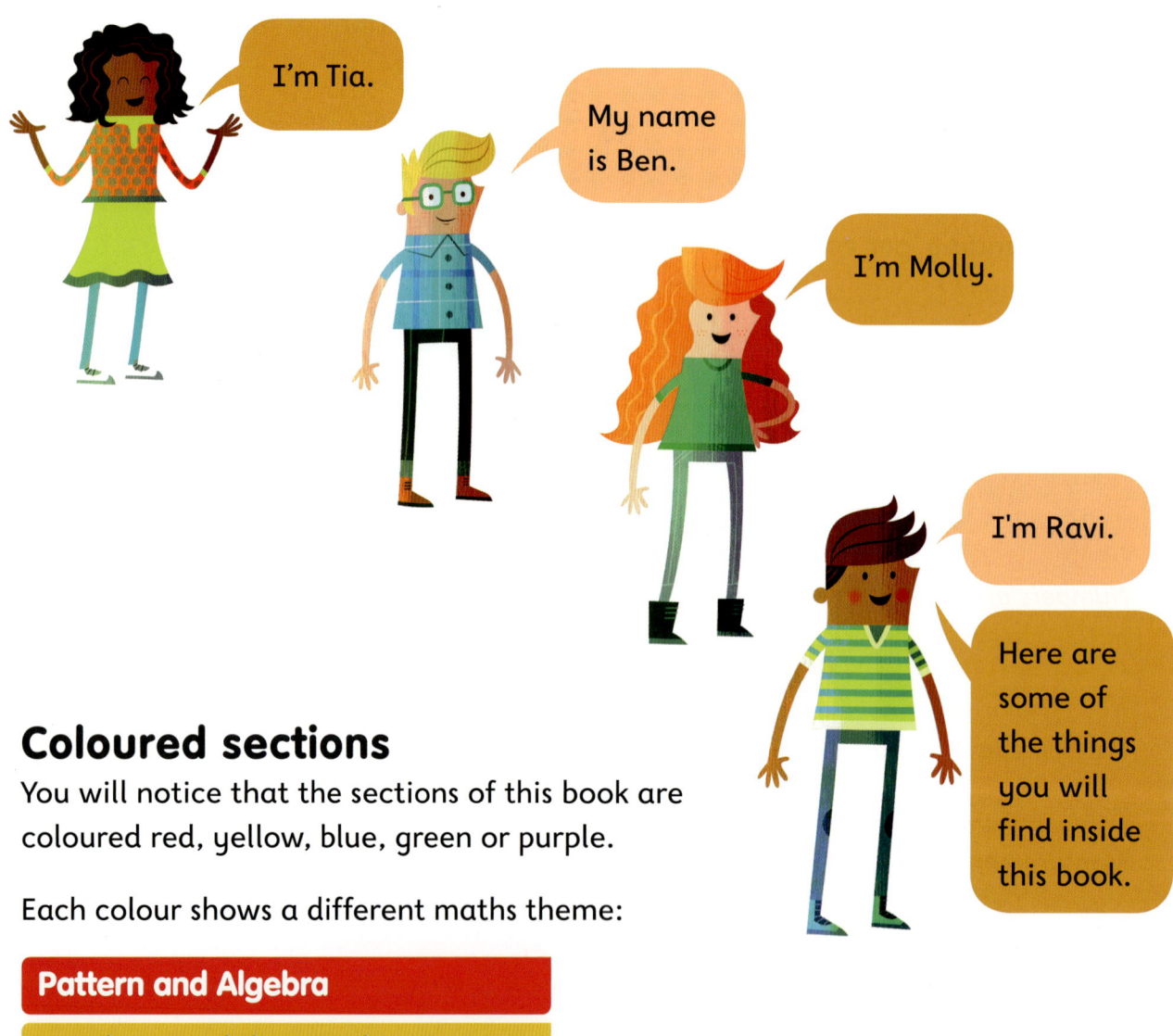

I'm Tia.

My name is Ben.

I'm Molly.

I'm Ravi.

Here are some of the things you will find inside this book.

Coloured sections

You will notice that the sections of this book are coloured red, yellow, blue, green or purple.

Each colour shows a different maths theme:

Pattern and Algebra

Numbers and the Number System

Calculating

Geometry

Measurement

 In this book you can try out new calculations…

 … and new methods for finding answers.

 There are opportunities to look for patterns…

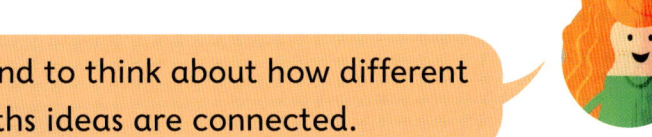 …and to think about how different maths ideas are connected.

Practice

These questions help you to practise and explore the new maths ideas you have learned.

Going deeper

These questions give you extra challenge and make you think deeply.

You will need to work with a partner on questions that have this symbol.

When you see this grey symbol you can do these activities in the Explorer Progress Book.

Glossary

There is a glossary of maths words in the back of the book. In the glossary you can look up the meaning of words you don't know.

Working with large numbers

Practice

1. Can you write down in words the number of people who are estimated to live in New York City?

 2. • Write down a 7-digit number, keeping it secret from your partner.

   • Read the number for your partner to write down.

   • Compare your numbers. Are they the same?

   • Swap over and play again.

   • Continue taking turns until you have each written ten numbers.

New York City Estimated population 8 550 405

Make sure you include some zeros in your numbers.

Going deeper

1. Using each of the digits below once, can you write down a number that lies between 3 286 471 and 4 183 762? Can you explain how you worked out your answer?

   | 1 | 2 | 3 | 4 | 6 | 7 | 8 |

2. Can you write a set of instructions for reading out 7-digit numbers? Try to write them so that a younger child could follow them.

Visualizing a million

I teaspoon = about 50 grains of rice

50 teaspoons = I cup

4 cups = I litre

Practice

1 Can you work out the approximate volume of one million grains of rice?

2 What other things can you think of that take up about the same amount of space (that is, have the same volume) as one million grains of rice?

Going deeper

1 Can you explain how long one million seconds is?

2 Can you work out whether you have reached a million days old?

3 Can you work out roughly how long ago you would need to have been born to be a million days old now?

4 How would you explain to a younger child how big a million is?

5 Can you think of some different ways of showing how big a million is?

Exploring place value

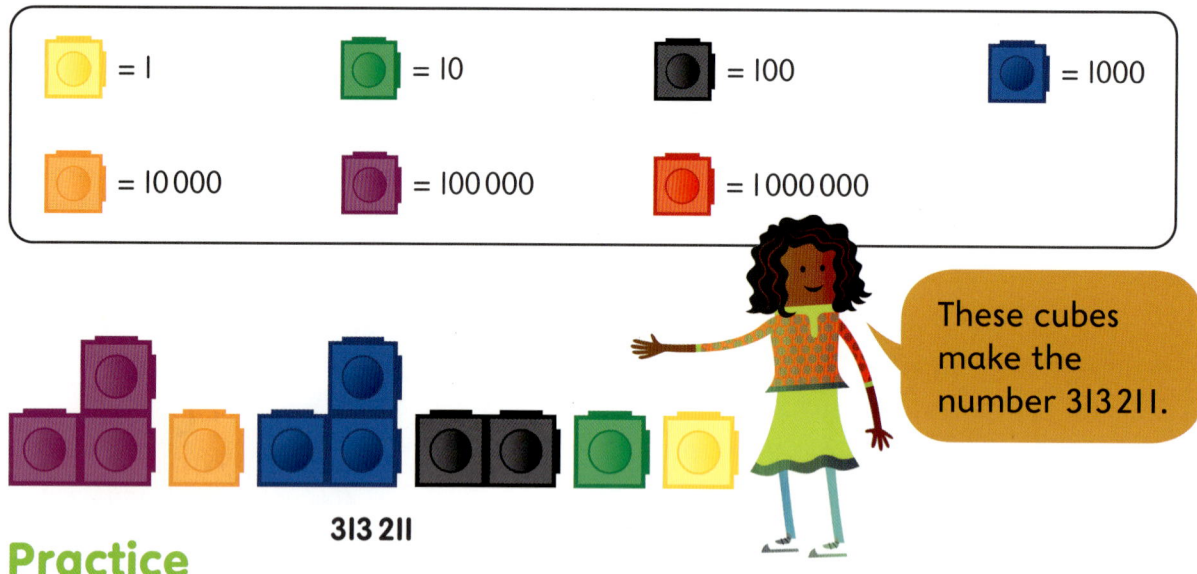

These cubes make the number 313 211.

313 211

Practice

1 Make a 6-digit number by rolling a dice six times (you can use the numbers you roll in any order). Then make your number with cubes.

Can you count on ten steps of 1000 from your number?

2 Make a 5-digit number by rolling a dice five times (you can use the numbers you roll in any order).

Can you count back from this number in steps of 100? What number will you arrive at if you count back ten steps of 100? Can you explain how you know?

3 Make five 7-digit or 6-digit numbers by rolling the dice and put all your numbers in order.

Going deeper

1 Take turns to make 7-digit numbers by rolling a dice five times and then including two zeros.

Can you subtract 90 000 from each number you make?

2 Can you change the game to make it even harder? What would be the hardest subtraction that you could think of? Can you make it easier?

Working with Roman numerals

In Roman numerals,
I stands for 1,
X stands for 10,
L stands for 50,
C stands for 100,
D stands for 500 and
M stands for 1000.

Practice

1 Can you work out what numbers these Roman numerals represent?

   a XXXVI
   b XLII
   c DCCXX

2 Can you use Roman numerals to write:

   a today's date
   b the date you were born
   c 4 July 1862?

Going deeper

1 a Which of these two numbers is bigger?
     Can you explain how you know?

   b Can you add these two numbers
     together?

   c Can you subtract the smaller number from the larger one?

2 Do you think it is easier to add and subtract with Roman numerals or
   with the numerals we usually use? Can you explain why, or why not?

3 What can you find out about how the ancient Romans actually did
   calculations?

Exploring equivalence with fractions

Practice

1 Can you find three different ways to describe the amount of these tiles that is coloured blue? Can you find:

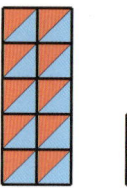

   a an improper fraction

   b a mixed number

   c a proper fraction?

2 a Can you draw an illustration that shows the fraction $\frac{18}{2}$?

   b Can you write $\frac{18}{2}$ in another way?

   c Can you explain how you worked this out?

3 Can you write these numbers in other ways?

   a $\frac{23}{2}$ b $12\frac{1}{2}$ c $\frac{38}{2}$ d $25\frac{1}{2}$

Going deeper

1 Can you draw tiles to show these fractions? You can use any design for the tiles you choose.

   a $\frac{1}{4}$ b $\frac{3}{4}$ c $2\frac{1}{2}$

2 Can you make an illustration that shows the fraction $\frac{9}{4}$?

3 a Can you explain how you did **question 2** above?

   b Can you use your method to make an illustration that shows $4\frac{3}{4}$?

4 a Which Numicon Shape would be most useful for illustrating quarters?

   b Can you explain why, and use some of the Shapes to illustrate $\frac{15}{4}$?

5 a Which number rod would be most useful for illustrating quarters?

   b Can you explain why, and use some of these rods to illustrate $2\frac{3}{4}$?

Converting mixed numbers and improper fractions

Practice

1 Can you write a fraction to go with this illustration?

2 How could you illustrate this same fraction with number rods?

3 a Try to find three ways of illustrating the fractions below.

$$\frac{13}{8} \qquad \frac{20}{8} \qquad \frac{31}{8} \qquad \frac{43}{8}$$

b What kind of fraction are these?

4 a Can you write the fractions you illustrated in **question 3** in another way?

b What kind of numbers have you written?

c Can you explain how you worked these out?

5 Which Numicon Shapes and which number rods could you use to illustrate the following numbers?

$$3\frac{1}{6} \qquad 4\frac{2}{3} \qquad 2\frac{3}{5} \qquad 1\frac{4}{7}$$

Going deeper

1 Can you write $3\frac{7}{8}$ in two other ways?

2 Explain how you could add together $4\frac{3}{5}$ and $\frac{19}{5}$.

3 Explain how you could subtract $\frac{19}{5}$ from $4\frac{3}{5}$.

Exploring equivalent fractions

The 12 children in Redwings group found these ways to divide their brownie tray into 12 equal parts.

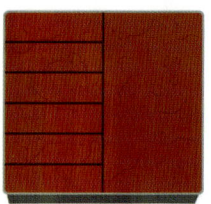 Dividing a half into sixths

 Dividing a third into quarters

 Dividing a quarter into thirds

Practice

1 There are 18 children in Kestrels group. How many different ways could they divide their brownie tray into 18 equal parts? You can use rectangles to record the ways you think of.

2 How many different fractions could you use to describe:

a $\frac{1}{3}$ of Kestrels' tray

b $\frac{1}{2}$ of Kestrels' tray

c $\frac{1}{6}$ of Kestrels' tray

d $\frac{1}{4}$ of Kestrels' tray?

3 How could you use number rods to show $\frac{1}{2}$, $\frac{1}{3}$, $\frac{1}{6}$ and $\frac{1}{9}$ of 18?

Going deeper

1 Can you use any other fractions to describe $\frac{1}{2}$ of anything? Can you explain this?

2 Could you use any other fractions to describe $\frac{2}{3}$ of anything?

3 Do any of the fractions below describe the same amounts?

$$\frac{2}{5} \qquad \frac{60}{100} \qquad \frac{20}{50} \qquad \frac{8}{10} \qquad \frac{9}{15} \qquad \frac{8}{20} \qquad \frac{6}{10} \qquad \frac{30}{50}$$

4 Can you solve these empty box problems and explain how you did this?

a $\frac{5}{8} = \frac{\blacksquare}{24}$

b $\frac{3}{9} = \frac{4}{\blacksquare}$

c $\frac{\blacksquare}{7} = \frac{6}{21}$

d $\frac{3}{\blacksquare} = \frac{12}{44}$

Using fractions in everyday situations

I make drinks with
I cup of squash
for every 5 cups of
water. The jug holds
12 cups altogether.

Practice

1 a Can you use number rods or Numicon Shapes to show the proportions of squash and water in Tia's favourite drink?

b Can you write these proportions as fractions?

2 Can you draw and fill in a table that shows how much squash and how much water is needed for 1, 2, 3, 4, 5 and 10 jugs of Tia's drink?

Number of jugs	Squash	Water
1	$\frac{2}{12}$	$\frac{10}{12}$
2	$\frac{4}{24}$	
3		

3 a What patterns can you find in the table you have made?

b Can you see any 'fraction families'?

Going deeper

1 Can you illustrate the squash fraction family using Numicon Shapes?

2 Can you illustrate the water fraction family using a number line?

3 Can you illustrate the $\frac{2}{5}$ fraction family using Numicon Shapes?

4 How can you test whether $\frac{3}{7}$ and $\frac{222}{518}$ are in the same fraction family?

Understanding decimals

This reading is between $37\frac{1}{4}$ and $37\frac{1}{2}$ degrees.

This one says it's 37·35°.

Practice

1 Does 37·35 lie between $37\frac{1}{4}$ and $37\frac{1}{2}$? Can you explain how you know?

2 How can you write 3 more decimals that lie between $37\frac{1}{4}$ and $37\frac{1}{2}$?

3 How can you write 37·75 as a fraction?

Going deeper

1 Can you explain why some numbers have a decimal point (for example 47·5) and others (for example 84) do not?

2 Can you describe where you have seen decimals being used, outside school?

3 Can you talk about any situations when you have used fractions, outside school?

Converting fractions to decimals

Practice

1 a How much of this baseboard is covered in blue?

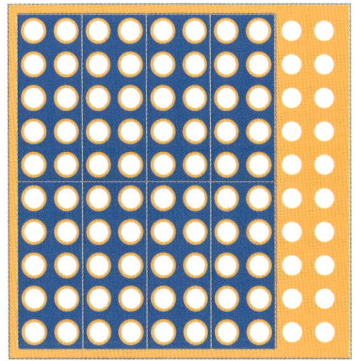

   b Can you write your answer as a fraction and a decimal?

2 Draw a number line like this.

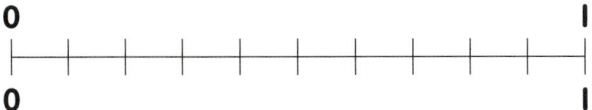

   Can you show your fraction and decimal from **question 1b** on it? How did you work out where these numbers go?

3 See if you can make a list of all the proper fractions that have denominators of 2, 4, 5, or 10.

   a Can you write them all down in order of size?

   b Can you write these fractions as decimals?

   c Now can you order the decimals on a blank number line?

Going deeper

1 a How many ways can you find to convert $12\frac{3}{4}$ to a decimal?

   b Which method do you prefer? Can you explain why to your partner?

2 Can you work any of your methods from **question 1** backwards? For example, what is 18·4 as a fraction?

3 Can you find a way to convert $\frac{2}{3}$ into a decimal?

4 Using a calculator, try investigating proper fractions that have denominators of 3 and 9. What happens when you convert these into decimals? Can you explain any of these results?

Thousandths as decimals

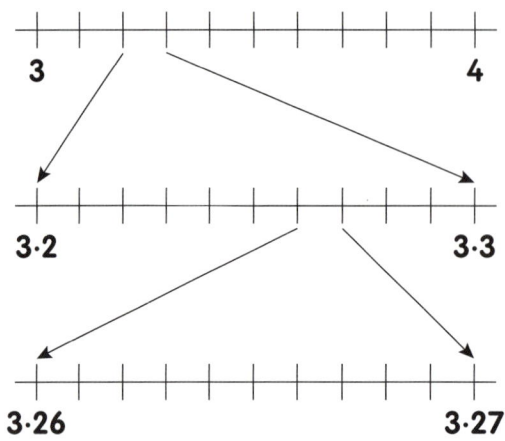

Practice

1 Can you explain where 3·264 would be on the diagram above?

2 Can you draw a set of number lines like these to show where you can find 2·586?

3 How could you write 3·264 as a fraction? Can you explain a good method for doing this?

Going deeper

1 How could you describe how far 0·586 is along a number line from zero?

2 Copy this empty decimal number line. Can you use it to show where you can find 5·289? How will you choose number labels for each end of the line?

3 Draw another empty decimal number line and label the ends '0' and '0·01'. Can you pick a point around the middle of the number line and say what number that is?

Comparing and ordering decimals

My file is 2·891 megabytes.

That's nothing! Mine is 2·879 megabytes.

Practice

1 Ravi and Tia can't agree whose computer file is bigger. One way of comparing two numbers like this is to use base-ten apparatus.

Does this apparatus help you to decide which number is bigger?

2 a Can you find another way to work out which of the numbers below is bigger than the other?

2·343 and 2·398

b Can you explain your method?

3 Four more children checked their document size in megabytes (MB) and these were: 2·81 MB, 3·147 MB, 3·009 MB and 2·908 MB. Can you put these files and Ravi and Tia's files above, in order of size? Which were bigger than 2·9 MB?

Going deeper

1 Which is bigger, $\frac{38}{100}$ or $\frac{364}{1000}$? Can you explain how you know? Can you write your answer using the signs '<' and '>'?

2 Which number is bigger, 0·02741 or 0·02729? Can you explain why you think so?

Measuring angles in degrees

That's an angle of about 30°.

I think it's more like 60°.

Practice

1 a Which estimate is closer, Ravi's or Molly's? How do you know?

b Can you say what kind of angle it is? How do you know?

2 Can you work out the size of each angle marked below, in degrees?

a **b** **c** **d**

3 a Can you estimate the size of these angles, by sketching each one?

| 90° | 45° | 30° | 120° | 20° | 70° | 160° | 225° |

b Can you explain how you made your estimates?

Going deeper

1 Can you work out what angle the **hour hand** of a clock will turn through, in degrees, between:

a 12 noon and 12:30 p.m.   **b** 12 midnight and 6:15 p.m.?

c Now compare the position of the **minute hands**. What is the angle between the minute hands in the two positions? Work this out for each pair of times.

Measuring angles with a protractor

Practice

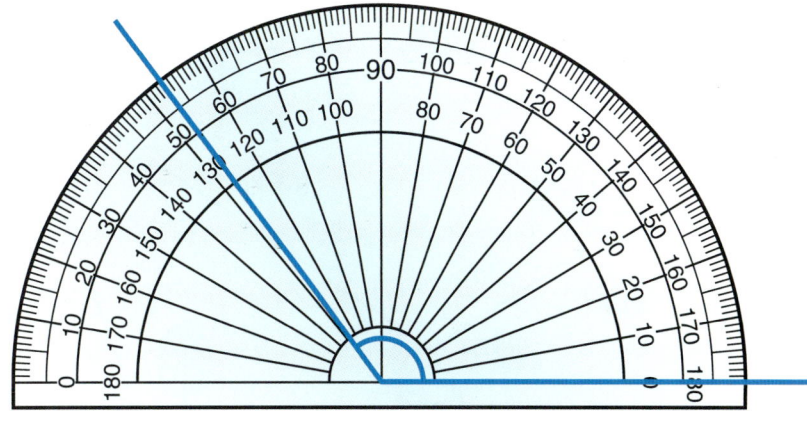

1 a Can you say what the angle shown is, in degrees?

b Can you explain how to measure an angle with a protractor?

2 For each angle, can you say what type of angle it is (acute or obtuse), estimate its size and then measure it with a protractor?

a

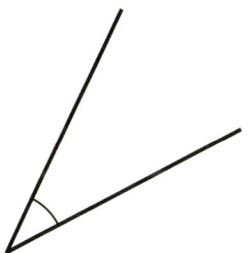

b

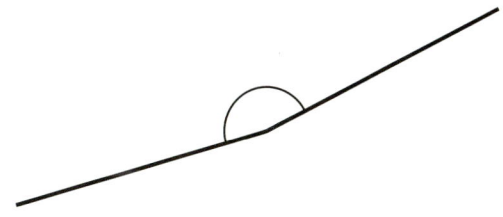

3 For each angle in **question 2**, sketch the angle, mark a reflex angle on your sketch and label it with its size. How did you work out these sizes?

Going deeper

1 a Adam is facing west. Can you say what bearing (angle clockwise from north) he is on?

b Adam now turns anti-clockwise to face north-north-east. Can you work out how many degrees he has turned through?

c Can you work out what his new bearing is?

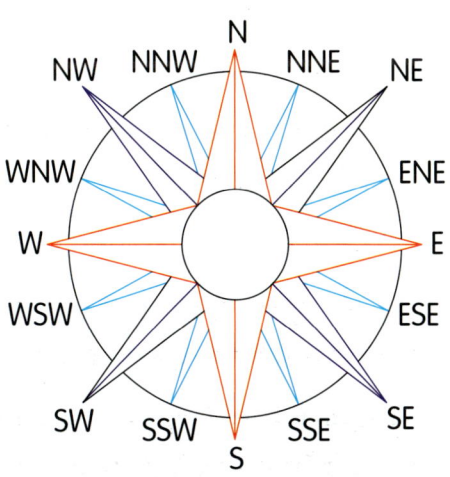

Measuring and drawing angles

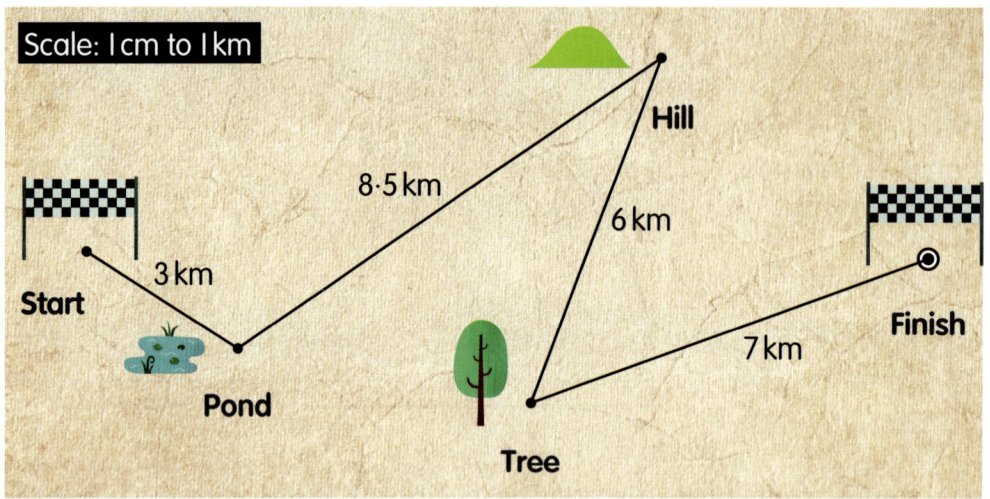

Practice

1 Ella is orienteering. She begins at the start, facing the pond. Can you use the map and a protractor to write a set of instructions for her to follow? Give the distances she should run and the turns she should make.

2 Can you draw two lines with an angle between them of:

a 45°   **b** 100°   **c** 108°?

....

Going deeper

1 • Draw a horizontal line 5 cm long using a ruler.

• Draw a second line, also 5 cm long, joined to the end of the first line. Use a protractor to make the angle between the lines 120°.

• Continue adding lines, each 5 cm long and joining at an angle of 120° to the previous line, until you end up back where you started.

• What shape have you drawn?

2 Repeat **question 1** but this time use an angle of 135°.
What shape have you drawn this time?

3 Compare the shapes you drew in **questions 1** and **2**. Can you explain what difference the size of the angle makes?

Finding missing angles

Practice

1 Tia is recording the view through her camera as it turns in a full circle.

   a How many degrees will the camera turn through to make a full circle?

   b If it has turned through 90° so far, how much further does the camera need to turn to complete a circle?

2 The diagram below shows a junction where a side road meets a main road at an angle of 55 degrees. Can you work out the sizes of the missing angles in the diagram? How do you know?

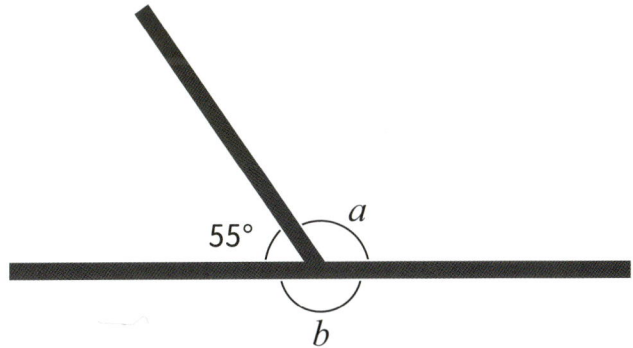

Going deeper

1 A security camera outside a museum is fixed on a pole and has been programmed to turn clockwise continuously. It pauses every time it points north-west, south-south-east and south-west. The camera's starting position is north-west.

What three angles does it turn through, between pauses? Can you draw a diagram showing these angles? How did you work out the answer?

Developing adding and subtracting

Super Swim
Target lengths: 65

Practice

1 Three children swam 65 lengths of a swimming pool in total between them. They each swam at least 18 lengths, but no more than 25 lengths. How many lengths might they each have swum?

2 Can you find five different solutions for each of these empty box problems?

   a ▨ + 580 + ▨ = 940   b ▨ + ▨ + 345 = 724

3 Can you solve each of these empty box problems?

   a 645 − 251 + ▨ = 646   b 58 + 76 = 53 + ▨

Going deeper

1 Can you explain why the number sentence 438 − 276 = 498 − 336 has to be true, without working out either subtraction calculation?

2 Look at your answer to **question 1**. Can you find nine other combinations of three whole numbers between 18 and 25 that total 65? You can use any of the numbers more than once. Do you think you have now found all the possibilities? Can you explain why?

3 What do you think is the quickest way to work out the following empty box problem: 89 − ▨ = 80 − 49?

Adding and subtracting in problem solving

Adding on from £3·25 to £10

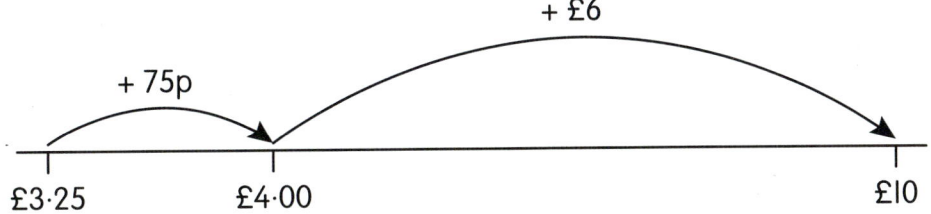

Subtracting £3·25 from £10

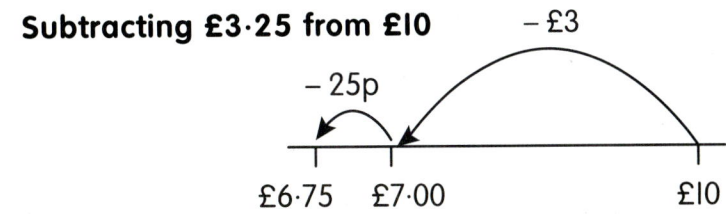

Practice

1 David's dad takes money out from the bank in £10 notes. If he bought items at the prices below with a £10 note each time, how much change would he get?

a £3·25

b £5·60

c £7·35

d £6·20

e £3·49

f £1·72

2 Alisha is doing a 90 km sponsored cycle ride from London to Brighton. She takes breaks at Kingston (after 23·1 km), at Horsham (after 52·8 km), and at Pease Pottage (after 70·4 km).

a When she is at Kingston, how far does she still have to go?

b When she is at Horsham, how far does she still have to go?

c When she is at Pease Pottage, how far does she still have to go?

Going deeper

1 If we know that two video clips together last 12 minutes 15 seconds, and that one of them lasts 5 minutes 54 seconds, what else can we work out? How many different number facts can you write out from this information?

2 What do you think are the best ways of doing the calculations for **question 1**? Can you explain why in each case?

Adding and subtracting fractions

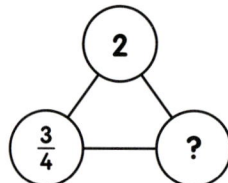

Practice

1 Can you complete the number trio above? Can you now write four different number sentences using just these numbers?

 2 Copy this blank number trio onto a whiteboard or into your exercise book. Write a whole number in the top circle and a fraction in one of the lower circles. Can your partner calculate what must go in the third circle? Swap roles and play again. Repeat four times.

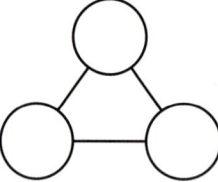

3 Can you work out what number goes in the empty box below?

$$3\frac{2}{5} + 2\frac{4}{5} = 3\frac{3}{5} + \blacksquare$$

Going deeper

1 What do you think are the best ways of calculating the answers to complete any number trios? Can you explain why?

2 What do you think is the best way of calculating the answer to this empty box problem? Can you explain why?

$$4\frac{5}{7} - \blacksquare = 2\frac{3}{7} - 1\frac{4}{7}$$

Adding and subtracting decimals

Practice

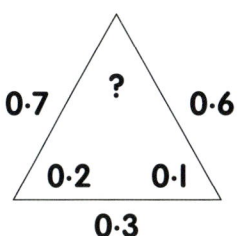

1 Can you find the missing number in this triangle?

2 Can you copy and complete these adding grids?

a

+	2·4	1·3
5·1		
4·7		

b

+		
	5·7	4·5
	10·1	8·9

3 Can you find another solution for the grid in **question 2b**?

Going deeper

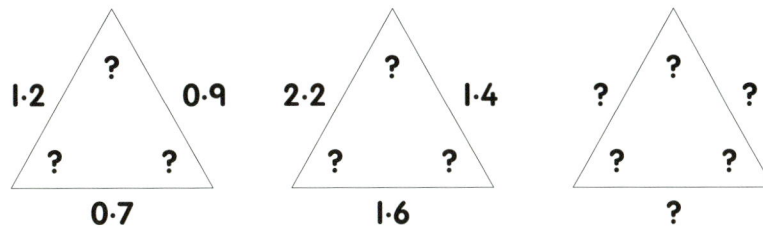

1 Can you suggest one way to complete each triangle above?

2 In a magic square, each row, column and diagonal adds up to the same total. Can you copy and complete this magic square?

0·7	0	
	0·4	0·6
0·3		0·1

3 Can you explain the quickest way to find the missing number below?

$$7\cdot35 - 3\cdot53 + \boxed{} = 7\cdot36$$

Using bridging strategies

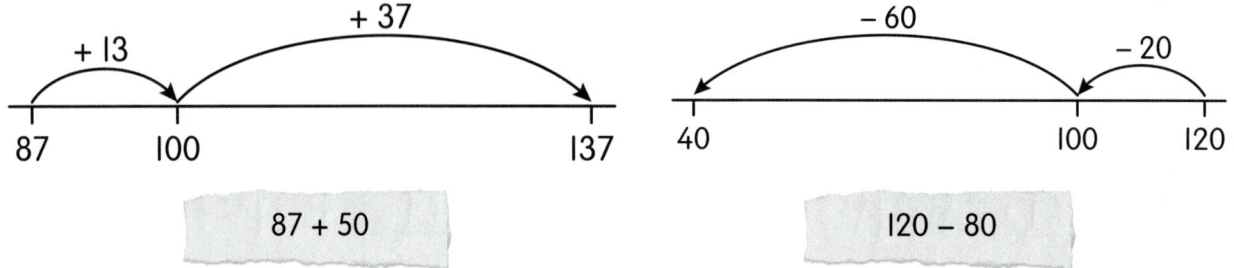

87 + 50

120 − 80

Practice

A: 170, 210, 340, 470, 520, 650, 6940, 7830

B: 47, 69, 85, 96, 389, 475, 580, 675

1 Choose four pairs of numbers, one from each list above, and either add them or subtract a number from List B from a larger number in List A.

2 Can you find one way to complete each of the calculations below?

a 195 + 75 = 195 + ▢ + ▢ = 200 + ▢

b 174 − 80 = 174 − ▢ − ▢ = 100 − ▢ = 94

c 360 − 85 = 360 − ▢ − ▢ = 300 − ▢ = 275

d 1600 + 540 = 1600 + ▢ + ▢ = 2000 + ▢ = ▢

Going deeper

1 80 + 57 = 87 + 50, but does 80 − 57 = 87 − 50? Can you explain?

2 What is the best way to work out how far you need to jump from any 2-digit number to 100?

3 Which methods would you choose to do the following calculations?

a 76 + 84 = ▢   b 154 + 132 = ▢   c 193 + 57 = ▢   d 293 + 198 = ▢

e 368 − 207 = ▢   f 240 − 122 = ▢   g 325 − 75 = ▢   h 476 − 97 = ▢

Bridging with time

Practice

1 Molly's favourite TV programme lasts 45 minutes. If it began at the following times, when would it end?

   a 2:00 p.m.   **b** 1:45 p.m.   **c** 3:30 p.m.

   d 4:25 p.m.   **e** 6:50 p.m.

2 Another programme lasts 50 minutes and ends at these times. What time did it begin?

   a 1:15 p.m.   **b** 5:30 p.m.   **c** 3:40 p.m.

   d 2:35 p.m.   **e** 8:10 a.m.

3 Naomi arrived back home at 4:10 p.m. after a journey of two and a quarter hours. When did she set off?

Going deeper

1 Jay has to get a meal ready for 1:30 p.m. The meat will take 1 hour 50 minutes to cook, the potatoes take 55 minutes, the carrots 40 minutes, the broccoli 6 minutes and the gravy 25 minutes. Can you make a list of the times when Jay has to start cooking each ingredient?

2 Tia's morning in school lasted 2 hours 45 minutes, and the afternoon was 1 hour 20 minutes long. She adds these times together using a column method and gets the answer 3 hours 65 minutes. Is she correct?

$$\begin{array}{r} 2:45 \\ +\ 1:20 \\ \hline 3:65 \end{array}$$

Bridging with fractions

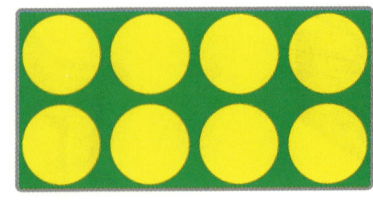

Practice

A: $3\frac{3}{8}$ $4\frac{1}{2}$ $2\frac{1}{4}$ $5\frac{1}{8}$

B: $\frac{5}{8}$ $\frac{3}{4}$ $\frac{7}{8}$

1 a Can you write and solve three different adding calculations by pairing a fraction from List A with a fraction from List B?

  b Can you write and solve three different subtracting calculations by pairing a fraction from List A with a fraction from List B?

2 Can you make two adding and subtracting questions with the answer $3\frac{2}{5}$?

Going deeper

1 What do you think is your best method for taking $\frac{7}{8}$ away from the mixed numbers in List A above? Can you explain why you prefer this method?

2 What is a good method for adding $\frac{3}{4}$ to the mixed numbers in List A?

3 What do you think is the best way of solving the problem below? Can you explain why?

$$\frac{5}{6} + \frac{13}{6} = 1 + \boxed{}$$

Bridging with decimals

Practice

1 Hazel has grown 0·8 cm during the Spring term, and she is now 131·3 cm tall. How tall was she at the beginning of term?

2 Write and solve three adding and three subtracting calculations by pairing a number from List A with a number from List B. (Avoid negative numbers.)

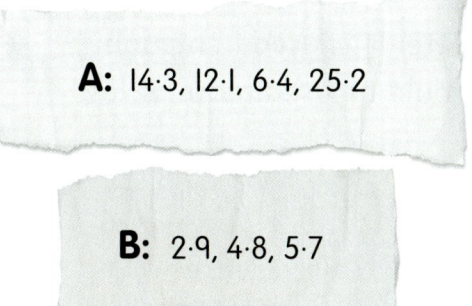

A: 14·3, 12·1, 6·4, 25·2

B: 2·9, 4·8, 5·7

3 Can you make two adding and subtracting questions with the answer 2·6?

Going deeper

1 What is the best method you can find for taking away 2·9 from the numbers in List A above?

2 What is a good method for adding 4·8 to the numbers in List A?

3 Can you find two different ways of solving this empty box problem? Does it help if you think about money? Can you explain?

7·6 − 2·99 = ▢

Everyday estimating

Practice

1 a Roughly how many times do you think the letter 'e' is used in a page of your reading book?

   b Approximately how many times do you think the letter 'e' is used in the whole book? How could you estimate this?

I have heard that the most common letter in the English language is 'e'. I'd like to test this.

2 Roughly how many times do you think the letter 't' is used in a page of your reading book? Do you think that it would be used more or less often than 'e'?

 3 How much do you think 1000 penny coins would weigh? Discuss how to estimate this with your partner.

A penny weighs 3·56 g.

Going deeper

1 What strategies would you recommend to estimate the number of tables in the school?

2 Chris is at a school fair, trying to guess how many sweets are in a big glass jar. Can you explain a good strategy for him to use to your partner?

Rounding to multiples of 10

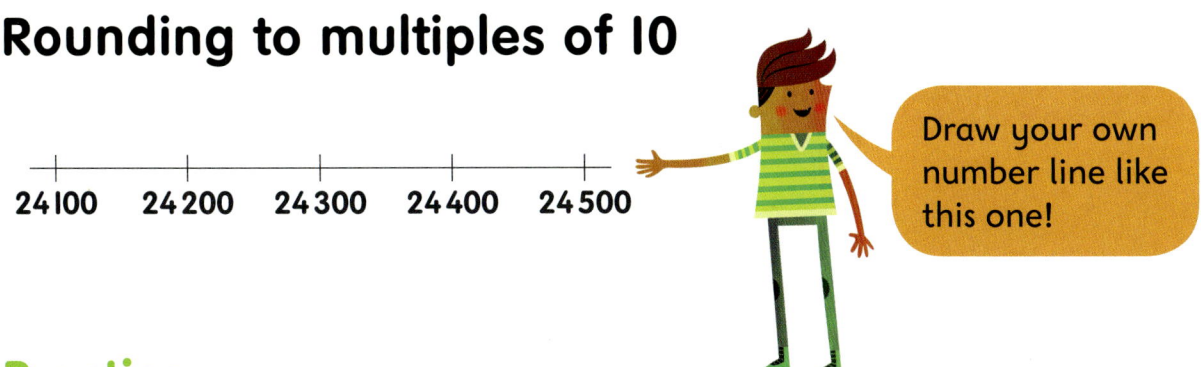

24 100 24 200 24 300 24 400 24 500

Draw your own number line like this one!

Practice

1 Can you show where you think 24 242 should be on your number line?

2 Rounding to the nearest 100, can you find one number that would round up and one that would round down to each entry on the number line? Make each of your numbers at least 30 away from the target.

3 Use scales to find the mass of one of your shoes, to the nearest 100 g.

4 Can you use a tape measure to measure the width of your table. How wide is it, to the nearest 10 mm?

5 The distance around the equator is 40 075 km. What is this to the nearest 1000 km?

6 Which of the numbers below is closest to 10 000?

10 030, 10 012, 9990, 9980, 10 019

Going deeper

 1 Can you round 1 738 562 to the nearest:

   a 10   **b** 100   **c** 1000

   d 10 000   **e** 100 000?

Can you explain the best way of rounding numbers like these?

2 The size of the crowd at a festival is 45 000 to the nearest hundred.

   a What is the smallest number that the crowd could be?

   b What is the largest number that the crowd could be?

Rounding with decimals

4·2 m

2·8 m

Practice

1 Gina wants to make the flower bed above. Can you work out what the area will be to the nearest square metre?

2 Seeds are bought in bags of 50 g, and should be spread at the rate of 2 g per square metre. Roughly how much of the bag will be left over after Gina has sown the seeds?

3 Gina wants to put edging around her flower bed, which comes in packs of 1·5 m for £3·69. How much edging do you think she will have to buy? How much do you think it will cost, to the nearest £1?

Going deeper

1 Josh said that he would rather be given a column of £2 coins as tall as himself than a collection of £1 coins weighing as much as himself. He is 140 cm tall and weighs 35 kg. A £2 coin is 2·5 mm thick and a £1 coin weighs 9·5 g.

Estimating roughly, do you think he made the right choice? Can you explain?

2 If you lined up 1 km of 10p coins, roughly what would their mass be?

A 10p coin has a diameter of 24·5 mm and a mass of 6·5 g.

Estimating calculations

2	3	5	7	8	9
120	210	290	420	490	630

Practice

 1 Look at the table above. Can you write a calculation using one number from the top row and one from the bottom row that will produce the answer closest to 150? Explain how you know you are correct.

2 Can you estimate the answers to the calculations below?

    a 4653 + 490 = ▨   **b** 8322 − 510 = ▨

    c 25 × 85 = ▨   **d** 3199 ÷ 38 = ▨

Going deeper

349, 303, 4509, 455, 798, 1003, 29, 12, 5, 20

1 Can you choose pairs of numbers from the list above to write calculations that will produce a number between:

    a 300 and 500   **b** 1000 and 1500   **c** 2000 and 5000?

2 Choosing any numbers from the list above, can you write a calculation that will produce the answer 600?

 3 Choosing pairs of numbers from the list above, can you write a calculation to produce:

    a the largest possible number

    b the smallest possible number?

Explain how you know you are correct to your partner.

Further adding and subtracting

456

I've used coloured cubes to illustrate the number 456.

	= 1
	= 10
	= 100
	= 1000
	= 10 000
	= 100 000
	= 1 000 000

Practice

1 Molly adds 1 black cube, 4 green cubes and 3 yellow cubes onto the number she made above. Can you write this as an adding calculation? What new total has she made?

2 Next she adds 4 yellow cubes. What is the new total?

3 Then she adds 4 black cubes. What is the new total?

4 Now she adds 1 purple, 3 orange and 2 blue cubes. What is the new total?

5 Now she takes away 10 000. Can you write this as a subtracting calculation? How much is left?

6 Finally, she takes away 4 green cubes. What is the final total?

Going deeper

1 If you start with 12 631 and you end up with 11 831, how much did you take away? How can you work this out in your head?

2 If you start with 133 003, which of the numbers in the list do you think are the hardest to take away? Can you explain why?

4, 40, 400, 4000, 40 000

Balancing with adding and subtracting

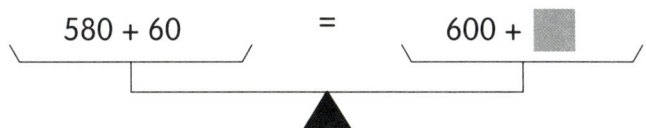

Practice

1 Can you find the missing number by balancing the calculation above?

2 Choose pairs of numbers from the list. For each pair, write one adding and one subtracting balancing calculation. For example, 379 − 169 = 380 − 170 , and 379 + 169 = 400 + 148.

379, 288, 169, 236, 187, 455

3 Can you solve the empty box problems below without working out the actual totals or differences?

a 695 + 40 = 700 + ▢   **b** 695 − 40 = 700 − ▢

4 Can you use balancing to make the calculations below easier?
For example, 287 + 63 = 300 + 50.

a 589 + 43 = ▢ + ▢   **b** 632 − 197 = ▢ − ▢

Going deeper

1 **a** Write an adding calculation that you can make easier using balancing. Now write an adding calculation that you think is **not** made easier using balancing.

b Can you explain the difference? Why does balancing help with the first calculation, but not the second?

2 When does balancing make subtracting easier? Can you explain why?

Adding and subtracting decimals by partitioning

Practice

1 Sam is cycling around this trail and starts at point A.

   a How far will Sam have cycled when he reaches each of the points B, C, D, E and F?

   b How far will he cycle altogether?

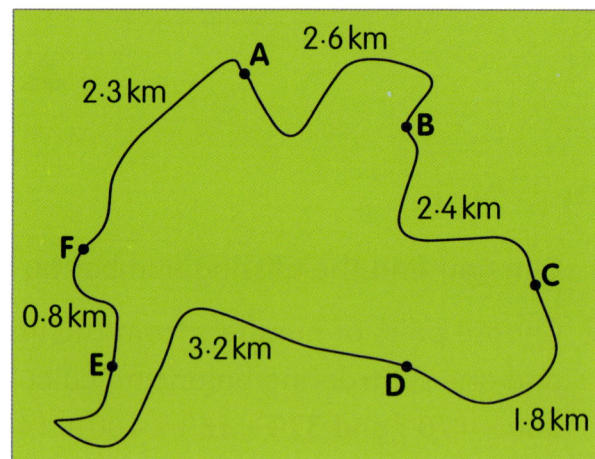

2 How many subtracting calculations can you write using one number from Spinner A and one number from Spinner B? What are the answers?

Spinner A

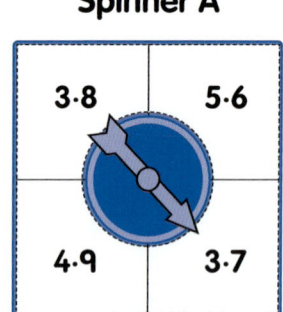

3·8	5·6
4·9	3·7

Spinner B

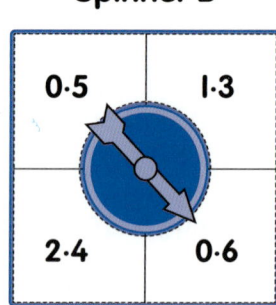

0·5	1·3
2·4	0·6

Going deeper

1 Can you explain which is your best method for:

   a adding 3·8 and 2·4

   b subtracting 2·4 from 3·8?

2 a Which way do you prefer to find the total of 5·2 and 2·4?

   b What about subtracting 2·4 from 5·2?

3 Can you explain anything about connections between the number pairs and the methods you prefer in **questions 1** and **2**?

Adding and subtracting in columns

Practice

1 Can you think of other ways of adding 167 and 156?

```
    1 6 7
  + 1 5 6
    3 2 3
      1 1
```

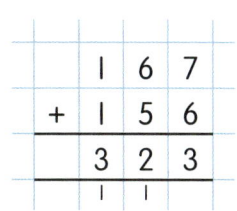

Hundreds	Tens	Ones

2 Can you think of other ways of subtracting 386 from 484? Which method do you think is best?

```
    ³4 ¹⁷8 ¹4
  -  3 8 6
       9 8
```

3 Lily and Ismail each chose a different item from the menu. The total was less than £13.

a Can you work out three different pairs of items they could have chosen?

b For each of these how much money would they have left if they started with £16·23?

Pizza	£9·85
Pasta	£6·99
Jacket potato	£3·45
Chicken	£4·49
Sandwich	£2·89
Fish fingers	£5·15

Going deeper

1 Can you solve these problems by using a written column method?

a 2326 + ▮ = 5798

b ▮ + 1786 = 4000

c ▮ − 4553 = 2345

d 7803 − ▮ = 4719

Sequences and patterns

Aanjay lives in Sumatra in Indonesia.
He has sent an email to the class.

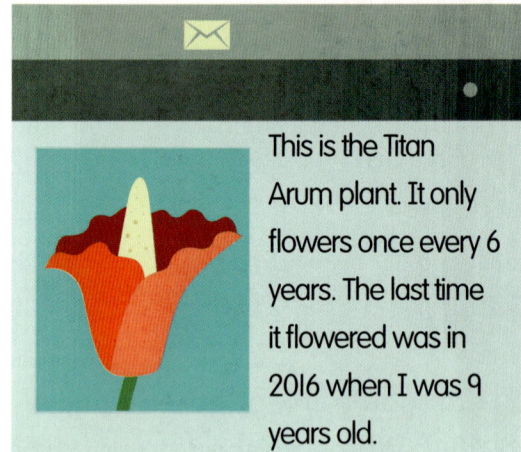

This is the Titan Arum plant. It only flowers once every 6 years. The last time it flowered was in 2016 when I was 9 years old.

Practice

1 Can you work out the years when Aanjay's Titan Arum flowered in the past 35 years? Write them down as a sequence.

2 Leap years happen every four years and 2016 was a leap year. Can you work out the next five years which will be leap years?

3 If you start with the year you were born, can you make a sequence by adding 9 years repeatedly?

4 Can you work out the missing numbers in this sequence?

3231, ☐, ☐, 3267, ☐, 3291

Going deeper

1 Can you make up another problem like **question 3** above, for a partner to solve?

2 If 123 is the first number of a regular sequence, and 168 is the last number, what are the possible numbers that could appear regularly spaced between these two?

3 How would you explain to a friend how to work out the answers to a question like **question 2** above? Can you make up a sequence problem of your own?

Decimal sequences

Practice

1 What number values could you give to these green and orange rods?

2 If the orange rods each had the value 'I', what value would you give to the green rods? Write these out as a sequence of decimal numbers.

3 Can you write a 0·3 sequence, starting at 0·5? Can you calculate when the first whole number will occur, before you reach it?

4 Can you carry on this sequence?

5·3, 5·7, 6·1, 6·5, ...

5 Try to make up the beginnings of some more decimal sequences for your partner to carry on with. You can make them as difficult as you like.

Going deeper

1 Jan made a different decimal sequence. Which numbers could go in these empty boxes?

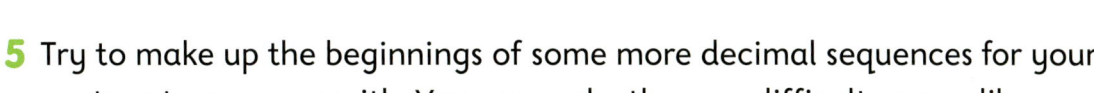

2 How could you use number rods to illustrate the sequence below? Can you explain how you worked out how to do this?

7·49, 7·56, 7·63, 7·7, 7·77, 7·84, ...

Fraction sequences

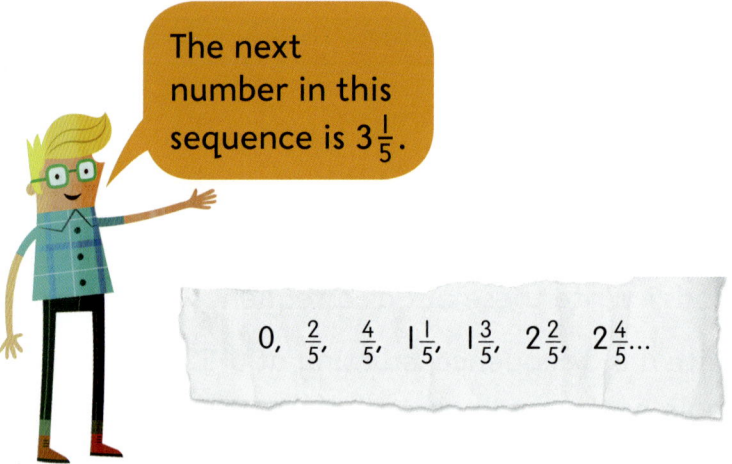

The next number in this sequence is $3\frac{1}{5}$.

$0, \ \frac{2}{5}, \ \frac{4}{5}, \ 1\frac{1}{5}, \ 1\frac{3}{5}, \ 2\frac{2}{5}, \ 2\frac{4}{5}\ldots$

Practice

1 How many different ways can you think of to illustrate the fraction $\frac{2}{5}$?

2 Can you use any of these different ways to illustrate the fraction sequence above?

3 Can you show these fractions as steps along a number line?

4 Do any of these illustrations help you to count on and back in fractions?

5 Can you continue counting on in the following sequences?

a $\frac{2}{3}, 1\frac{1}{3}, \ldots$ b $\frac{5}{6}, 1\frac{2}{3}, \ldots$ c $\frac{3}{8}, \frac{3}{4}, \ldots$

Going deeper

1 a Can you work out what the tenth term of the following sequence will be, without working out the first nine terms?

$0, \frac{7}{9}, 1\frac{5}{9}, \ldots$

 b How do you know your tenth term is correct? Can you check? Can you explain to a partner how you worked this out?

Connecting sequences with equivalent fractions

$$\frac{2}{3}, \quad \frac{4}{6}, \quad \frac{6}{9}, \quad \frac{8}{12} \dots$$

Practice

1 Can you carry on writing out this sequence of fractions?

2 What do you think is the easiest way to continue writing out this sequence?

3 What can you say about **all** of these fractions?

4 Try writing out fraction families, in the easiest way that you know, for each of these fractions:

a $\frac{1}{3}$ b $\frac{2}{7}$ c $\frac{5}{8}$ d $\frac{3}{5}$

Going deeper

1 How do we know that lists like these really do show fractions of the same families? How can you test them?

2 Can you think of other ways to generate fraction family lists?

3 If you were explaining what fraction families are to a friend, how could you use a number line to help illustrate what you are saying?

4 Could you use Numicon Shapes to explain why different members of a fraction family are definitely in the same family?

5 Is there a quick way to work out the tenth member of the $\frac{3}{8}$ family, without working out the first nine members as well? Can you explain your method?

Transformations

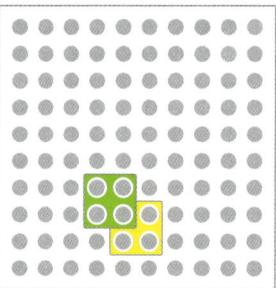

Practice

1 How many symmetrical shapes can you make on a baseboard using just one yellow and one light green Numicon Shape?

2 How many ways can you find to add a third Numicon Shape to the board to create a symmetrical figure? You can use **any** of the Numicon Shapes.

 3 Mia wants to use a symmetrical pattern of tiles in her new bathroom. Can you create a design for the tiles using three types of Numicon Shapes on a baseboard? Swap with your partner, and find the line of symmetry on each other's design.

Going deeper

1 How many symmetrical figures can be created by using two 2-shapes and two 3-shapes on a 5 × 5 section of a baseboard? Here is one possible arrangement:

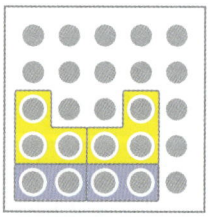

 2 Take turns to draw a symmetrical pattern onto a folded piece of squared paper, without your partner seeing. Can you describe the pattern in words for your partner to draw? Afterwards, show them the original pattern to check.

Reflections on a coordinate grid

Practice

1 Copy the coordinate grid with the triangle shown here. Can you reflect the triangle in three different ways, using each of the triangle's sides as a line of symmetry?

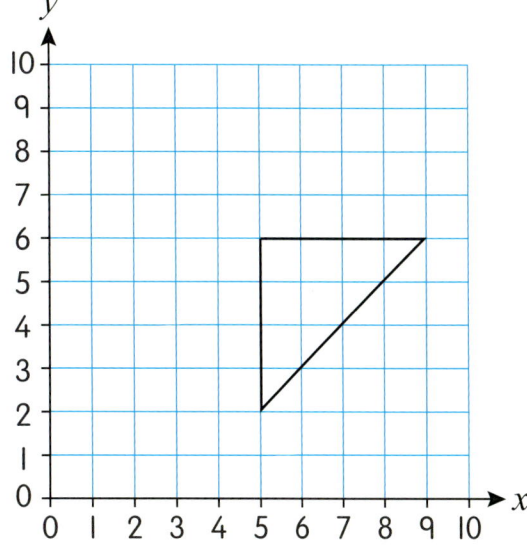

2 What is the same about all of the new triangles you have drawn? What is different? Explain to a partner.

3 If you think of the four triangles as a single shape and draw around the outline, what shape have you drawn? Can you find its line of symmetry?

Going deeper

1 Take turns to choose a shape and plot it onto a coordinate grid. Give your partner the shape's coordinates and choose a line for them to reflect the shape in. Ask your partner to plot the reflected image.

2 The 'Clear Reflections' window cleaning company is looking for ideas for a logo using their initials (CR) and two different reflections.

Using a coordinate grid with axes from 0 to 10, and the letters C and R, create a symmetrical design for the logo. List the coordinates for your design.

Describing translations using coordinates

I drew a rectangle with vertices (1,8), (1,10), (4,10) and (4,8).

I drew a rectangle with vertices (10,7), (10,9), (7,9) and (7,7).

I drew a rectangle with vertices (4,4), (7,4), (7,2) and (4,2).

I drew a rectangle with vertices (1,2), (4,2), (4,4) and (1,4).

 Ben

 Molly

 Ravi

 Tia

Practice

1 a Copy rectangle A on a coordinate grid (with axes from 0 to 10). Mark on the rectangle for each child.

b How would you translate rectangle A on to each of the children's rectangles?

2 How would you translate each of the children's rectangles onto rectangle A?

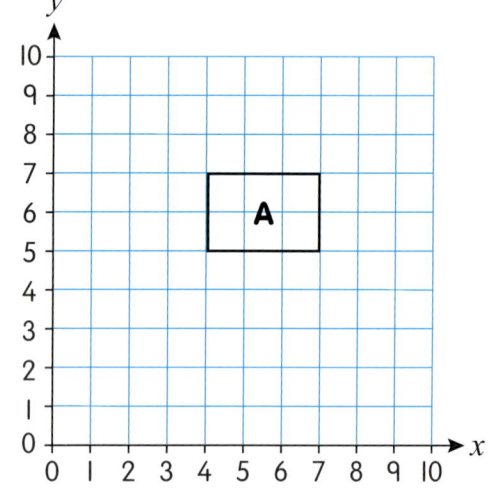

Going deeper

1 If you combine three of the translations listed below, one of the children's rectangles above will translate onto rectangle A.
Can you work out whose rectangle it is?

| Right 4, down 2 | Left 16, up 3 |
| Right 10, down 7 | Right 6, up 7 |

2 Can you describe any reflections you can see in your diagram from **question 1**?

Exploring translations on a coordinate grid

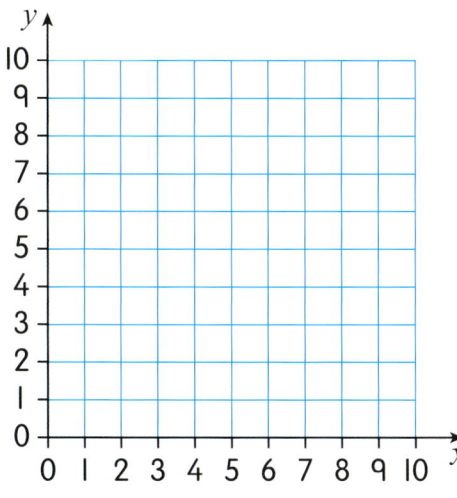

Triangle A has coordinates (2,6), (2,4) and (5,4)

Triangle B has coordinates (7,6), (7,4) and (10,4)

Triangle C has coordinates (1,9), (1,7) and (4,7)

Triangle D has coordinates (3,10), (3,8) and (6,8)

Triangle E has coordinates (1,3), (1,1) and (4,1)

Triangle F has coordinates (7,10), (7,8) and (10,8)

Practice

1 Draw the triangles on a coordinate grid like the one above.

2 Find out which triangle would map onto which other triangle using these translations:

a 5 right

b 6 up

c 2 right, 1 up

d 4 left

e 6 right, 3 up

f 6 left, 3 down

g 6 left, 7 down

Going deeper

1 Where would you draw the coordinates of a new triangle (G) so that each of the translations below would map it to one of triangles A–F?

2 left	1 left, 3 up
7 up	4 right, 3 up

2 Draw a design on a coordinate grid and give your partner instructions for a translation. Can your partner plot the new position of the design? How can you check their answer? Swap roles and repeat three times.

Working with negative numbers

Practice

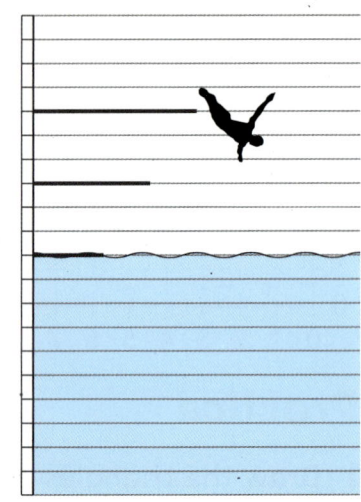

1 Which depths do you think are further from the surface of the water:

a ⁻6 m or ⁻8 m

b ⁻3 m or ⁻2 m

c ⁻5 m or ⁻4 m?

2 As a diver dives, what happens to the numbers describing their height above the water? Can you explain why this is?

3 As a diver just touches the water, what height are they at? What depth are they at? Can you explain?

4 As a diver goes down into the water, what happens to the numbers describing their depth? Can you explain why this is?

Going deeper

1 The peak of Mount Everest is described as 8872 metres 'above sea level'. Can you explain why its height is described in this way?

2 The greatest depth of the Atlantic Ocean, at Milwaukee Deep, is 8350 metres below sea level. If the sea level were to rise by one metre, how deep would Milwaukee Deep be then?

3 The Greek philosopher Aristotle was born in 384 BCE and lived for about 60 years. Roughly when do you think he died? Was it around 440 BCE, or around 320 BCE? Can you explain your answer using a timeline?

Negative and positive temperatures

Practice

1 Can you put the following temperatures in order, from coldest to warmest?

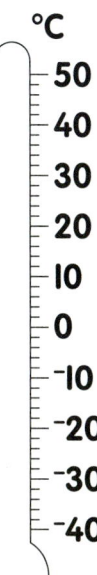

11°C, 2°C, ⁻7°C, ⁻9°C, 1°C, ⁻3°C, 4°C

2 Can you count aloud in degrees from 5°C to ⁻5°C? Now write down the temperatures that you named, in order.

3 Can you put these negative temperatures in order, starting with the coldest?

⁻17°C, ⁻6°C, ⁻9°C, ⁻11°C, ⁻3°C, ⁻32°C, ⁻21°C, ⁻14°C

Going deeper

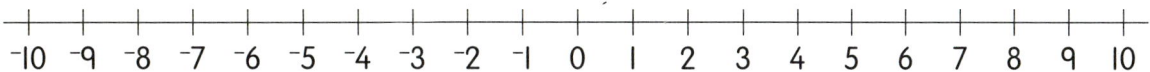

1 Draw a number line, like the one above. Colour the numbers on your line that show temperatures hotter than ⁻3°C in red and circle numbers that show temperatures colder than 1°C in blue.

2 Look at the number line below. Can you find the missing numbers for **a–c**?

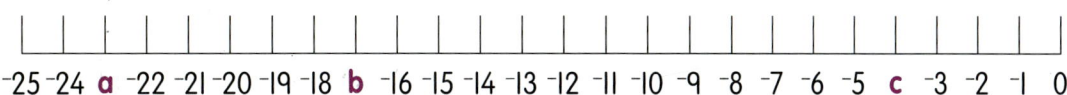

3 When measuring temperatures we sometimes use negative numbers. Can you think of any other measures that use negative numbers?

Temperature differences

Highest recorded temperatures

Place	Highest temperature
Death Valley (USA)	57 °C
Kharga (Egypt)	50 °C
Athens (Greece)	48 °C
Sadovo (Bulgaria)	45 °C
Marlborough (New Zealand)	42 °C

Lowest recorded temperatures

Place	Lowest temperature
Yukon (Canada)	⁻63 °C
Lapland (Finland)	⁻52 °C
Luhansk (Ukraine)	⁻42 °C
Uyuni (Bolivia)	⁻26 °C
New South Wales (Australia)	⁻23 °C

Practice

1 Can you calculate the temperature differences between:

   a Death Valley and Yukon   **b** Athens and Luhansk

   c Yukon and Luhansk   **d** Marlborough and Uyuni

   e Lapland and New South Wales?

2 Which of the places above are hotter than 46 °C? Which of them are colder than ⁻40 °C?

3 Can you write the temperature that is 5 °C warmer than each of the lowest recorded temperatures in the table above?

4 Can you write down four pairs of numbers, one positive and the other negative, that each have a difference of 12 °C?

Going deeper

 1 Which method would you use to calculate the difference between 18 °C and ⁻5 °C? Can you explain and show this using a number line?

2 Can you explain why the difference between ⁻9 °C and ⁻15 °C is a positive number?

Negative numbers and direction

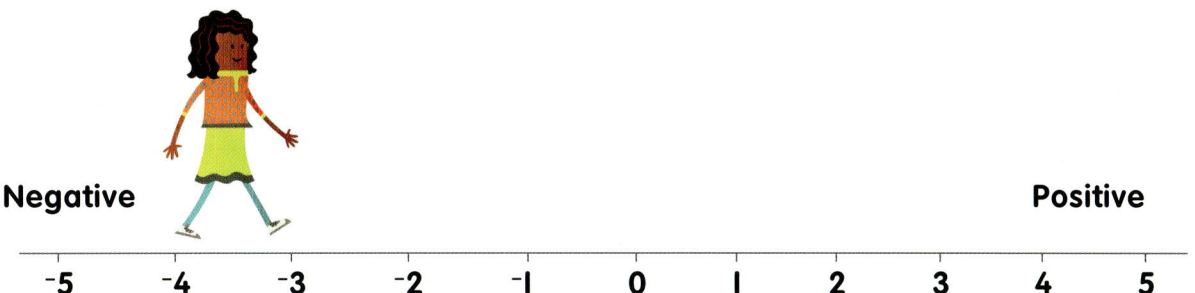

Negative   Positive

`-5   -4   -3   -2   -1   0   1   2   3   4   5`

Practice

1 If Tia starts at ⁻4 and takes 7 steps in a positive direction, where do you think she will be standing now?

2 If Jo starts at 2 and takes 9 steps in a negative direction, where do you think she will be standing now?

3 Ravi starts at 11 and jumps 3 steps at a time in a negative direction. Can you work out which numbers he lands on with his first 10 jumps?

4 Steph starts at ⁻14 and jumps 4 steps at a time in a positive direction. Can you work out which numbers she lands on with her first 10 jumps?

Going deeper

1 If you start at ⁻4 and take 7 steps in a positive direction, you land on 3. If you start at 4 and take 7 steps in a negative direction, you land on ⁻3. Will this always work? Can you explain, using a number line?

2 Can you complete the sequence below by working out what should go in the empty boxes?

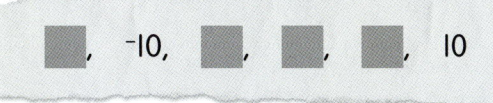

▉, ⁻10, ▉, ▉, ▉, 10

3 Can you complete this sequence by working out what should go in the empty boxes?

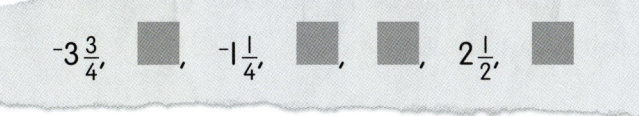

$-3\frac{3}{4}$, ▉, $-1\frac{1}{4}$, ▉, ▉, $2\frac{1}{2}$, ▉

Developing multiplying and dividing

Two Numicon Shapes are missing!

Practice

1 All the different products of pairs of the four Numicon Shapes above are: 8, 10, 18, 20, 36 and 45. Can you work out what the two missing Shapes are?

2 Can you work out which four Shapes can be paired to give the products 30, 35, 40, 42, 48 and 56?

3 Which four numbers could be paired to give the products 63, 77, 84, 99, 108 and 132?

Going deeper

1 Can you explain why there are usually six different products in the questions above?

2 Sometimes there are fewer than six different products. Can you explain why?

3 Is it possible to choose four numbers that give only **even** numbers as products? Can you explain why, or why not?

4 Is it possible to choose four numbers that give only **odd** numbers as products? Can you explain why, or why not?

Short written methods: multiplying and dividing

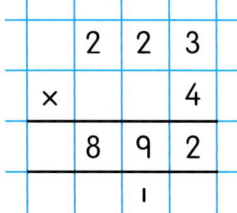

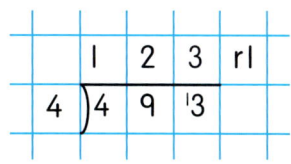

Practice

1 Can you arrange the Numeral Cards below to make HTO × O calculations that give the greatest and the smallest products?

2   2   3   4

2 Can you use the same Numeral Cards to make HTO ÷ O calculations that give the greatest and the smallest quotients?

3 Can you work out the missing digits?

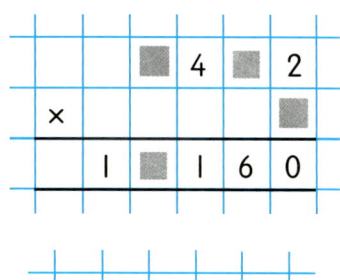

4 Can you work out the missing digits?
Could there be more than one answer?

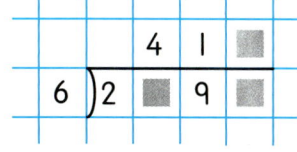

5 Can you make up some more problems like **questions 3** and **4** for your partner to solve?

Going deeper

1 What strategies did you use for answering **questions 1** and **2** above? Will your strategies always work? Can you explain?

2 When are written ways of multiplying and dividing the best methods to use? Can you explain, using some examples?

Fractions of an amount

We could easily colour in $\frac{1}{2}$ of this oblong exactly.

Practice

1 Which other fractions of the oblong would also be easy to colour in exactly? Can you explain why?

2 Can you write down $\frac{1}{7}$ of each of the numbers in the list shown here? Try to draw an array of squares like the one above to show that at least one of your answers is correct.

14, 35, 84, 56, 21, 63

3 Which multiplying and dividing facts would you use to solve these:

a $\frac{1}{3}$ of 27   b $\frac{1}{6}$ of 42   c $\frac{1}{8}$ of 72   d $\frac{1}{4}$ of 64?

Going deeper

1 a 30 is $\frac{1}{4}$ of what?   b 76 is $\frac{1}{3}$ of what?

c What fraction of 108 is 12?   d What fraction of 72 is 18?

2 Look at this pattern. What do you notice? Can you continue the pattern?

$\frac{1}{3}$ of 24 = 8

$\frac{1}{6}$ of 24 = 4

$\frac{1}{12}$ of 24 = 2

3 Can you make up two more patterns like the one in **question 2**?

4 Using the array at the top of the page, can you find any more easy fractions to colour in exactly? Can you give some examples of fractions that would be very difficult to colour in exactly?

Multiplying and dividing decimals

3 × 0·6 = 1·8
and
1·8 ÷ 3 = 0·6

Practice

1 For three mornings each week, Eva has to walk 0·4 km to school. How far does she walk to school altogether each week? Can you write this as a calculation?

2 James has to walk to school on four mornings each week, and altogether he walks 2·8 km in one week. How far does he have to walk to school in the morning? Can you write this as a calculation?

3 Can you multiply each of the following numbers by 3, 5 and 6?

   a 0·2   b 0·3   c 0·5   d 0·6   e 0·8   f 0·9

4 Can you divide each of the following numbers by **two** whole numbers less than 10 (for example 1·5 ÷ 3 = 0·5, 1·5 ÷ 6 = 0·25)?

   a 1·5   b 2·4   c 3·6   d 4·2   e 5·2

Going deeper

1 Knowing that 4 × 0·7 = 2·8, can you write out **two** related dividing calculations?

2 Knowing that 2·4 ÷ 4 = 0·6, can you write out **four** related multiplying calculations?

3 Can you work out 13·2 ÷ 11? Which tables facts can help you to do this?

Using proper fractions

$\frac{1}{3}$ lemon juice

$\frac{1}{2}$ lemon juice

Practice

1 One jug of lemonade was made using $\frac{1}{3}$ lemon juice, and the rest was water. Another jug was made using $\frac{1}{2}$ lemon juice, and the rest was water.

   a Which do you think tasted stronger and why?

   b How much of the weaker lemonade was water?

2 Which fraction is bigger, $\frac{3}{8}$ or $\frac{1}{2}$? Can you use number rods, a number line, or Numicon Shapes to show why?

3 Which fraction is bigger, $\frac{2}{3}$ or $\frac{5}{6}$? Can you show why?

4 Which fraction is bigger, $\frac{6}{10}$ or $\frac{12}{20}$? Can you explain, or show, why?

Going deeper

1 Amy played three pieces of music in a competition, and she scored: 4 out of 5, 7 out of 10, and 17 out of 20. Which was her best score? Can you explain why?

2 Would you rather be right 56 times out of 64, 27 times out of 32, or 3 times out of 4? Can you explain why?

3 Can you put the fractions below in order of size, starting with the smallest?

$$\frac{2}{3} \qquad \frac{73}{108} \qquad \frac{49}{72} \qquad \frac{7}{12} \qquad \frac{25}{36}$$

4 Can you write three fractions where the three denominators have a common factor? Can you now put them in order of size?

Common denominators

Molly plays twice in every nine beats.

Ravi plays four times in every six beats.

Tia plays one beat in every two beats.

Practice

1 If the piece of music is 36 beats long, who will have played their instrument the most times? Who will have played the fewest times? Can you explain the quickest way to work out these answers?

2 Choose three of the fractions below and put them in order of their size, beginning with the smallest.

$$\frac{2}{3} \qquad \frac{3}{8} \qquad \frac{3}{4} \qquad \frac{5}{9} \qquad \frac{7}{12} \qquad \frac{8}{15}$$

3 Choose a different set of three fractions from the same list and put them in order of size, beginning with the largest. Can you show these in position on a number line?

Going deeper

1 Which of the fractions in the list above is closest to 1?

2 Look at the list of fractions above.
   a How many different sets of three fractions can you make from the list? How do you know you have found all the possible sets?

   b Which set of three fractions gives the highest total when you add them together?

3 Dev has dug $\frac{1}{4}$ of a vegetable patch, Sarah has dug $\frac{1}{3}$ and Clara has dug $\frac{3}{8}$ of the patch. Who has dug the most? Who has dug the least? How much of the vegetable patch still has to be dug?

Using 'greater than' and 'less than' signs

$$\frac{7}{36} \quad \frac{5}{6} \quad \frac{3}{4} \quad \frac{2}{3} \quad \frac{1}{2} \quad \frac{5}{9} \quad \frac{7}{12} \quad \frac{11}{18}$$

Practice

1 Choose five pairs of fractions from the list and show which is the smaller fraction in each pair by using the '<' sign.

2 Choose five different pairs of fractions from the list, and show which fraction is greater in each pair by using the '>' sign.

 3 Can you put all the fractions in the list in order, beginning with the smallest? Can you explain to your partner how you did this?

Going deeper

1 Using the numbers 5 and 9 only once, can you make this statement true?

$$\frac{\square}{18} > \frac{\square}{12}$$

2 Using the numbers 5, 6 and 9 only once, can you make this statement true?
Is there another way to do this?

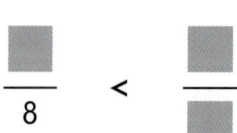

Simplifying proper fractions

Practice

1 The lemonade above was made from 15 cups of lemon juice and 25 cups of water. What proportion of the lemonade is lemon juice and what proportion is water? Can you simplify your fractions to their lowest terms?

2 Can you write six different fractions that are all equivalent to $\frac{3}{5}$?

3 Simplify each of the fractions below to its lowest terms.

$$\frac{24}{32} \qquad \frac{16}{64} \qquad \frac{12}{18} \qquad \frac{30}{36} \qquad \frac{84}{96} \qquad \frac{60}{108} \qquad \frac{44}{55} \qquad \frac{65}{120} \qquad \frac{49}{56} \qquad \frac{48}{51}$$

Going deeper

1 Can you simplify $\frac{48}{53}$? Can you explain your answer?

2 What is the closest fraction to $\frac{48}{53}$ that you could simplify? Can you explain why you think this is the closest fraction?

3 What advice would you give to someone who wants to know how to simplify fractions in the quickest way?

Inverse relationships

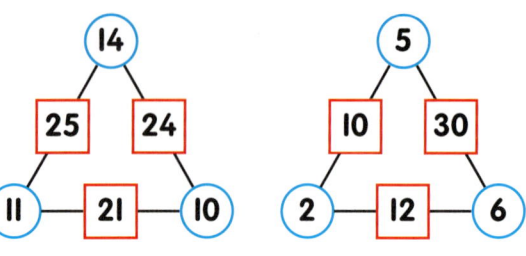

Practice

1 Can you explain the relationships between the numbers in the circles and the numbers in the squares, in these arithmagons?

2 Can you work out the missing numbers in these arithmagons?

a

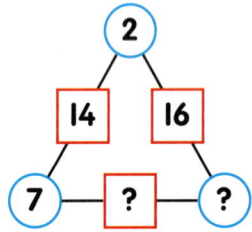

b

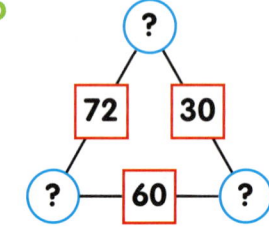

3 Look at the numbers below. Can you put these together in an adding arithmagon?

'Circle' numbers: $\frac{3}{4}$ $1\frac{1}{4}$ $1\frac{1}{2}$

'Square' numbers: 2 $2\frac{1}{4}$ $2\frac{3}{4}$

··

Going deeper

1 What is your best method for solving arithmagon problems like **question 2b** above? Can you explain clearly, for a friend to understand?

2 Can you make an arithmagon that uses 0·5 as a circle number, and 2·7 and 1·3 as square numbers?

3 Can you make a hexagonal arithmagon that uses six circle numbers, for your partner to solve? How many squares and/or circles can you leave blank if there is to be only one solution? Which ones?

4 Can you make triangular arithmagons that use these sets of numbers?

A: 5, 8, 12, 40, 60, 96

B: 7, 11, 18, 18, 25, 29

Missing number calculations (+ and −)

Practice

1 Can you explain a good strategy for working out the missing numbers in this calculation?

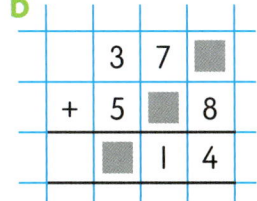

```
    2  9  ■
+   4  ■  5
─────────────
    7  4  3
```

2 Can you copy and complete the empty box calculations below?

a ■ + 75 = 90

d 120 − ■ = 51

b
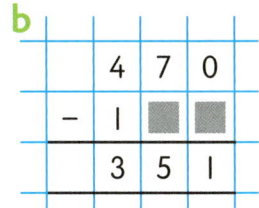

```
    3  7  ■
+   5  ■  8
─────────────
    ■  1  4
```

c
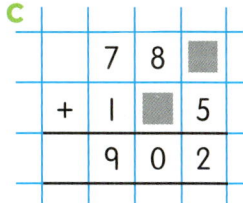

```
    5  1  3
−   2  ■  ■
─────────────
    2  2  8
```

3 Now try to copy and complete these empty box calculations.

a
```
    6  3  ■
+   1  ■  5
─────────────
    8  3  1
```

b
```
    4  7  0
−   1  ■  ■
─────────────
    3  5  1
```

c
```
    7  8  ■
+   1  ■  5
─────────────
    9  0  2
```

 4 Can you make up some new problems like this for your partner?

5 Try making this problem into a column subtraction and then solve it.

3456 − ■■■■ = 678

Going deeper

1 Work out the missing numbers. How can you check that the whole calculation is correct in **two** different ways?

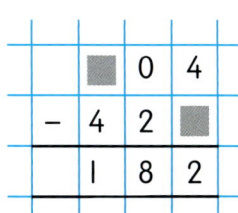

```
    ■  0  4
−   4  2  ■
─────────────
    1  8  2
```

2 Can you find **two** different ways to solve this missing number problem?

5774 = ■■■■ − 3185

Missing number calculations (× and ÷)

	■	7	■
×			5
	8	7	0

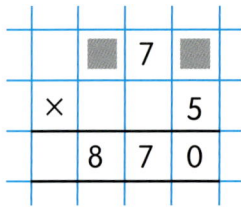

Practice

1 Can you explain a good strategy for working out the missing numbers in the calculation above?

2 Can you work out the two dividing problems below?

		■	2
8)	3	3	■

		6	■
7)	4	■	5

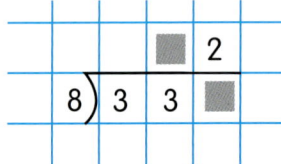

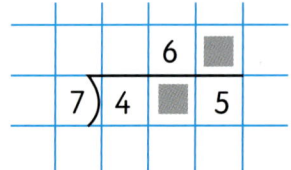

3 Can you make up some new problems like **question 2** for your partner to solve?

Going deeper

1 Copy and complete the empty box calculation below. How can you check that the whole calculation is correct in **two** different ways?

	2	■	1	■
×				3
	6	0	4	8

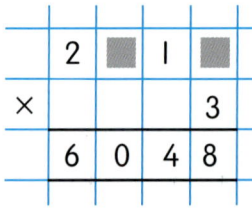

2 Can you find **two** different ways to solve the empty box problem below?

1586 = ■ ■ ■ ■ ÷ 6

Solving problems by working backwards

I bought a comic for £1·99, five snack bars for 52p each and a pencil case for £2·99. I've still got £2·42 left from my week's pocket money!

I know how much you get every week.

Practice

1 How does Tia know how much pocket money Ben gets every week?

2 What if Ben had £4·42 of his pocket money left? Can you explain what his weekly pocket money would be then?

3 Gemma spent half of her pocket money on a ball and had £2·50 left. Can you work out her weekly pocket money?

Going deeper

1 If you thought of a number, multiplied it by 5, added 6, divided by 7 and got the answer 8, what number would you have started with? Can you explain how you worked this out?

2 Why will the answer to the problem on the right always be 5?

3 Agata and Harry have £74·20 between them, but Agata has £4·80 more than Harry. Can you calculate how much they each have?

Think of a number.

Multiply it by 5.

Add 25.

Divide it by 5.

Take away the number you first thought of.

Mental and written methods of adding

Four classes have been collecting school vouchers.

	March	April
Blue Class	2330	1760
Green Class	1438	2376
Red Class	1695	2245
Orange Class	2834	1967

Practice

1 What is your preferred method for working out the total vouchers collected by each class? Can you explain why and then find the totals?

2 Look at the numbers in the list below. Which pairs of numbers can you add together mentally? Are you sure you have found them all?

> 2025, 1598, 2320, 3165, 2756, 4775, 4670, 2438

Going deeper

1 Look back at **question 1**. Choose one of the totals you worked out.

   a Can you describe **two** different ways of checking your answer?

   b Do you think one of the methods you have described is better than the other? Can you explain why or why not?

2 Blue and Green Classes above are both Year 5, and Red and Orange Classes are Year 6. Without calculating these totals exactly, can you estimate whether Year 5 or Year 6 collected the most vouchers? Can you explain how you estimated these totals?

Using the written column method for adding

Practice

1 The crowd at a tennis match is made up of 7453 children and 8278 adults. The crowd at a football match has 7546 children and 8154 adults. Can you work out which match has the bigger crowd?

2 Rearrange the four digits 6, 4, 3 and 9. Which pair of 4-digit numbers can you make that together:

a have the closest possible total to 8000

b have the closest possible total to 9000?

Check your answers to a and b by using the written column method shown in the example below.

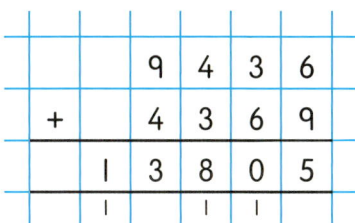

Going deeper

1 Use the digits 2, 5, 6 and 8 once each time to make different 4-digit whole numbers. How many different pairs of numbers can you make that total less than 8000? Can you explain how you know that you have found all the possible pairs?

2 I rearrange the numeral cards 3, 4, 6 and 9 to make two different 4-digit whole numbers. The total of my two numbers is 8018.

Can you use the column method for adding, to find my two numbers?

Adding money

Natural Yoghurt
£1·24

Oats
£1·67

Peanut Butter
£2·78

Apple Juice
£2·42

Rice
£2·56

Raisins
£3·25

Practice

1 Marek bought a bottle of apple juice and one other item. He only had £5 to spend. Which pairs of items could he have bought? Write out your calculations using a column method.

2 Sally bought three items. The total cost was less than £6.
Which groups of three items could she have bought?
Have you found all the possibilities? Can you explain how you know?

Going deeper

All the price labels in a shop can be made using these cards:

£ · 3 2 5 0 9

For each price, each card can be used only once. Not all the cards need to be used each time.

1 Use the cards to make pairs of prices with a total between £18 and £18·50.

How many pairs can you find? Have you found all the possibilities? Can you explain how you know?

2 The cards are used to make two prices with a total of £12·84. Can you work out what the prices are?

Adding decimals and measures

Distances between towns in kilometres				
	Ashford	Brighton	Chichester	Dorking	Eastbourne
Ashford		83·81	102·87	61·82	48·29
Brighton			32·19	37·05	24·09
Chichester				43·89	53·38
Dorking					57·62
Eastbourne					

Practice

1 Guna is driving from Ashford to Brighton, and then from Brighton to Chichester. Would her journey be longer or shorter if she travelled directly from Ashford to Chichester?

2 If you travel from Chichester to Dorking and then on to Eastbourne, how far will you travel in total?

3 Use the table to find journeys that are less than 90 km long and call in at three towns. Have you found them all? How do you know?

Going deeper

1

2·478 kg 3·188 kg 4·095 kg 5·209 kg

a Two of the parcels together weigh 6573 g. Can you say which two?

b On some balance scales, two parcels are placed on each side, so that the scales are as close as possible to balancing. Which two parcels are on each side?

c How much do all four parcels weigh, in total?

d Do you think it is easier to do calculations like these in grams or in kilograms, or is there no difference? Why do you think this?

Mental and written methods of subtracting

		$^4\cancel{5}$	$^{16}\cancel{7}$	$^1 7$	
	−		3	8	9
			1	8	8

Hundreds	Tens	Ones

Practice

 1 Can you explain to a partner how the written column method of subtracting shown above works?

2 Choose a number from each list below and subtract B from A. How can you check your answer? Were you right?

A: 904, 812, 722, 931, 821, 715

B: 189, 276, 391, 469, 586, 678

3 Repeat the steps of **question 2** for six more subtracting calculations, checking your answers each time.

Going deeper

1 Choose one of your subtracting calculations from above. Is there another method you could use to do this calculation? Which method do you prefer and why?

2 a How many different subtracting calculations could you make by subtracting a number in List B from a number in List A?

b Which of these calculations would you find easier to do another way?

3 Do you have a favourite method for subtracting? Can you explain why, using some of the examples above?

Using the written column method for subtracting

Practice

1 What is the difference in length between the longest and the shortest rivers in this table? How could you check your answer?

2 Can you find out which pair of rivers has a difference in length that is closest to 500 miles?

River	Length in miles
Amazon	4345
Nile	4258
Yangtze	3917
Mississippi	3902
Yellow River	3395
Mekong	2705
Volga	2266
Indus	1976
Danube	1795
Ganges	1628
Rhine	768
Loire	629

Going deeper

1 Using the digits 1, 3, 4 and 5 once in each number, which pair of 4-digit numbers can you subtract to give an answer as close as possible to 2000?

2 Choosing pairs of numbers from the table above, which differences between them would you find easiest to calculate mentally?

Which differences would you prefer to use a written column method of subtracting for?

Can you explain why?

Subtracting money

£27·49

£52·10

Practice

1 How much more expensive is the second table lamp than the first?

2 Over six weeks, Emily's family spent the following amounts on their weekly supermarket shop:

Week I	Week 2	Week 3	Week 4	Week 5	Week 6
£82·17	£96·05	£92·57	£115·26	£87·63	£78·83

a What is the difference between their most expensive and their least expensive shopping bills?

b Can you work out which was the biggest difference between two consecutive weekly shopping bills?

Going deeper

1 The difference between two of Emily's family's weekly shopping bills was £13·74. Can you work out which two bills these were, using the table above?

2 Can you work out the difference between the total cost of the supermarket shopping above for the first three weeks, and the total for the second three weeks?

3 Try to work out the subtraction on the right in two different ways: mentally, and as a written column method. Which method did you find easier? Can you explain why?

£20·00 – £13·68

Subtracting decimals and measures

Practice

1   In March, Angela's delivery journey to Paris used 40·352 litres of petrol, but when she came back she only used 36·687 litres. How much less petrol did she use on the way back?

2   Other journeys Angela made in April used this much petrol:

5·438 litres	22·641 litres	15·507 litres
8·007 litres	10·206 litres	18·889 litres

   a   What was the biggest difference in litres used between any two of her journeys in April? Can you explain how you worked this out?

   b   What was the smallest difference between any two of Angela's journeys in April?

Going deeper

1   Can you work out which two amounts of petrol in **question 2** have a difference that is closest to 9 litres? How do you know?

2   Two other journeys used 84·46 litres, and 75·57 litres. How many different ways can you find to calculate the difference between these volumes? Which method do you prefer and why?

3   Which was the hardest calculation you had to do on this page? Can you explain why? Can you find any way to make it easier?

Multiplying and dividing by 10

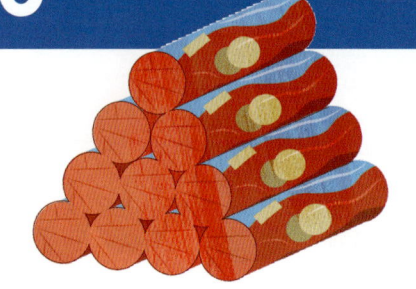

There are 15 biscuits in a packet and 10 packets of biscuits.

Practice

1 How many biscuits are there altogether? Can you show this with base-ten apparatus and a place value chart?

2 There are 10 packets of each of the following biscuits. How many of each biscuit will there be altogether?

a
8 biscuits in a pack

b
24 biscuits in a pack

c
35 biscuits in a pack

3 If a packet of 10 biscuits weighs 160 g, approximately how much does one biscuit weigh?

4 Can you work out the following?

a $215\text{g} \div 10$ b $£1·50 \div 10$ c $1\text{kg} \div 10$

Going deeper

1 a A packet of biscuits costs 70p. How much would 10 packets cost? Give your answer in pounds and pence.

b If 10 packets of biscuits cost £12, how much would one packet cost? Give your answer in pounds and pence.

2 Rosa says, "When you multiply by 10, you add a zero and when you divide by 10, you take a zero off." Her teacher says this is incorrect. Can you say why, using examples, and explain what the correct rule is?

Multiplying and dividing by 100 and 1000

One pack of 150 rounders balls

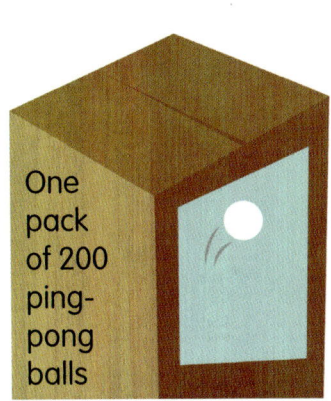
One pack of 200 ping-pong balls

One pack of 100 tennis balls

Practice

1 A sport shop buys 100 of each of these packs of sports balls. How many balls of each type will they have?

2 a The pack of tennis balls above weighs 2700 g. How much must one tennis ball weigh, approximately?

b A larger pack of 1000 ping-pong balls weighs 600 g. Can you work out how much one ping-pong ball must weigh, approximately?

3 Can you work out each of these calculations?

a 5 kg ÷ 1000   b £1500 ÷ 1000   c £1·20 × 100   d 2500 g ÷ 100

Going deeper

1 If 1000 rounders balls cost £1720 and a single ball weighs 74 g, can you work out the total weight of the balls and the cost per ball?

2 A shop wants 1000 tennis balls and 2000 ping-pong balls. 100 tennis balls cost £150 and 200 ping-pong balls cost £230. Can you explain to the shop how to work out the total cost? What is the total cost?

Multiplying and dividing decimals by 10

I whole

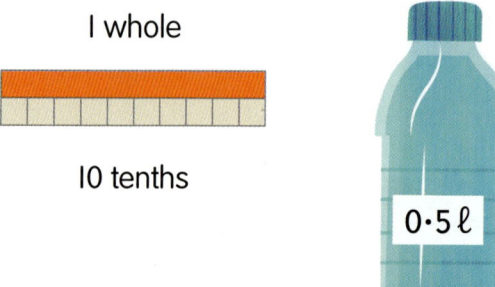

10 tenths

0·5 ℓ

Practice

1 If one bottle of water holds 0·5 ℓ, how much water is there in 10 bottles? Can you represent this with number rods?

2 Can you multiply these decimal numbers by 10?

   a 4·6   b 0·7   c 0·35   d 0·62

3 A carton of orange juice contains 2·2 ℓ. If we want to share this among 10 cups, approximately how much would there be in each cup? Give your answer in millilitres.

Going deeper

1 This risotto recipe serves ten people.

   Can you say how much of each ingredient you will need for one person?

0·75 kg risotto rice

1·5 ℓ hot vegetable stock

0·6 kg peas

0·25 kg mushrooms

 2 a Can you explain how you would work out 3·2 ÷ 10?

   b Can you work out 2·4 ÷ 10?

3 Which of these calculations do you think are correct? Explain why or why not, and what the correct answer should be.

   a 0·9 ÷ 10 = 9   b 0·3 ÷ 10 = 0·03

   c 12 ÷ 10 = 1·2   d 7 ÷ 10 = 0·07

Multiplying and dividing decimals by 100 and 1000

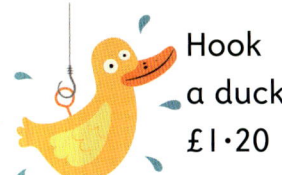 Hook a duck £1·20

 Bouncy castle £1·50

 Raffle £2·50

 Coconut shy £1·80

The price of a ticket for each activity at a fair is shown above.

Practice

 1 How much will each stall make if they sell the number of tickets below?
Can you explain your method using apparatus of your choice?

    a 100 tickets   **b** 1000 tickets

2 Can you calculate the following?

    a $317 \div 100$ **b** $2005 \div 1000$ **c** $45·8 \times 100$

3 Can you explain how to get from 540 to 0·54?

4 What do you need to do to get from 1030 to 10·3?

Going deeper

1 Can you find the missing number in this sequence?

    4500, ?, 0·45, 0·0045

2 Can you explain what is happening in this sequence?

    3·2, 32, 0·32, 320

3 If you know $1·5 \times 100 = 150$, can you write down five other related facts?

4 Take turns to play this game with a partner.

- Write down a number, then multiply or divide it by 10, 100 or 1000.
- Write down the new number and ask your partner to work out the operation you used.

Exploring different units of measurement

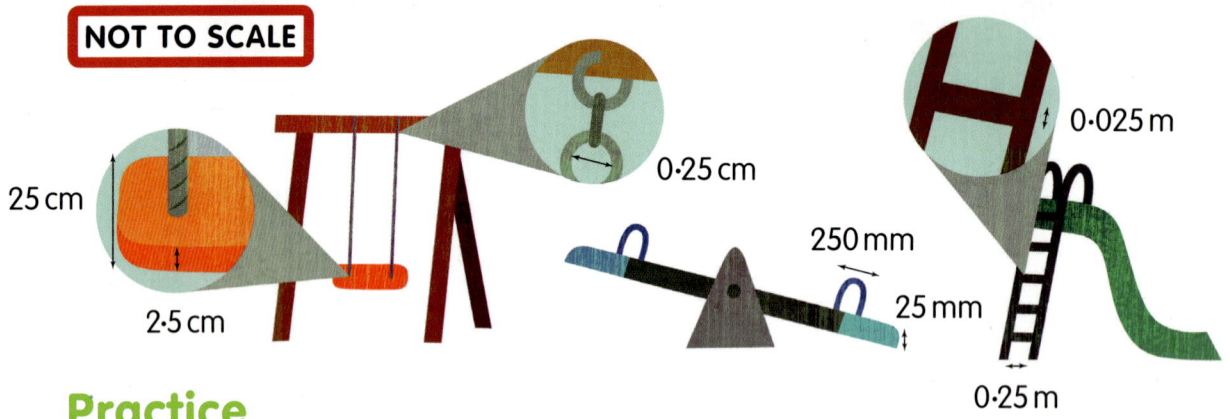

NOT TO SCALE

25 cm

2·5 cm

0·25 cm

250 mm

25 mm

0·25 m

0·025 m

Practice

1 Can you find sets of equivalent measurements from this playground?

2 Which measurement is the odd one out? Can you explain why?

Going deeper

1 The children all tried to grow the tallest sunflower to plant in the border of their playground. Who won – and by how much?

Tia: 2·6 m

Ben: 2 m 50 cm

Ravi: 248 cm

Molly: 0·003 km

2 The table shows the jumps three athletes made in a long jump competition. Only their longest jump counts.

	1st Jump	2nd Jump	3rd Jump
Max	1·58 m	1·6 m	1·49 m
Jenny	147 cm	181 cm	177 cm
Ali	1 m 45 cm	1 m 67 cm	1 m 47 cm

Can you work out who came 1st, 2nd and 3rd?

Exploring metric and imperial units

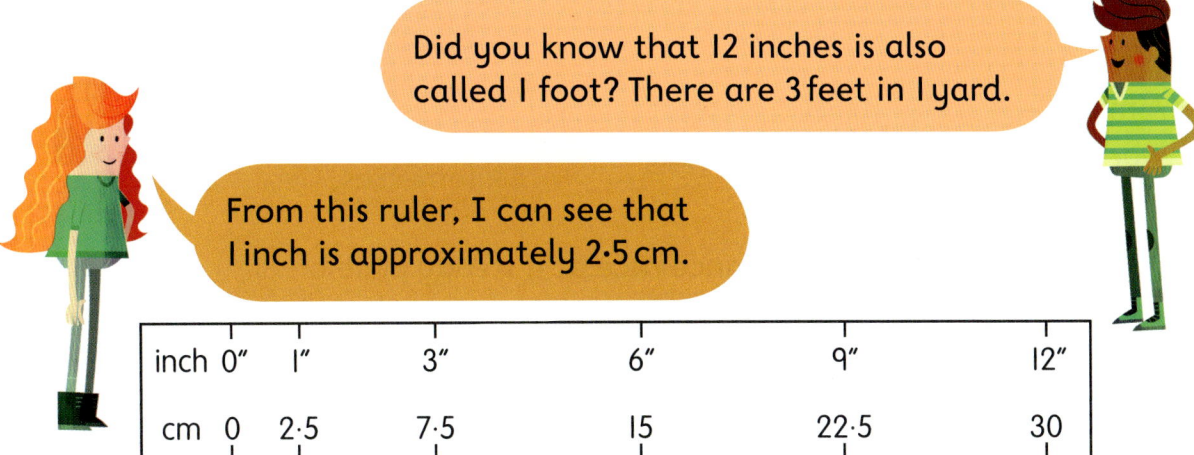

Did you know that 12 inches is also called 1 foot? There are 3 feet in 1 yard.

From this ruler, I can see that 1 inch is approximately 2·5 cm.

inch	0"	1"	3"	6"	9"	12"
cm	0	2·5	7·5	15	22·5	30

Practice

1 Find metric approximations for these distances:

   a 6 feet   b 12 yards   c 120 yards   d 30 yards

2 Find imperial approximations for these distances:

   a 30 cm   b 1 m   c 45 cm   d 100 mm

Going deeper

1 Using Molly's scale above, can you work out who is likely to be right?

Our fence is the tallest. It's 7 feet tall.

I think ours is taller – it's 215 cm tall.

2 Mel is hungry. Can you help her by working out which shop is nearer? How much nearer it is?

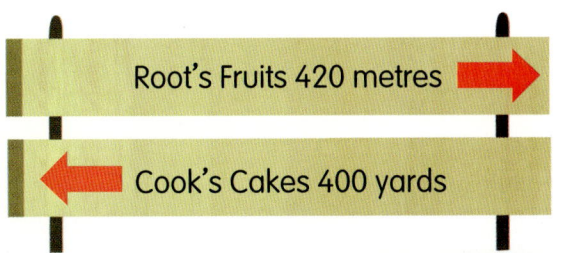

Root's Fruits 420 metres

Cook's Cakes 400 yards

I yard is about 90 cm.

71

Converting between miles and kilometres

Practice

1 A marathon is 26·2 miles. Can you work out approximately how many kilometres this is?

5 miles is approximately equal to 8 kilometres.

2 If Tia's mum runs a long-distance race at a steady pace of 9 minutes per mile, how many kilometres will she cover in the first hour?

Going deeper

1 A car drives at a steady speed on a motorway and covers 70 miles per hour. Can you use what Tia says above to work out approximately how long it will take to drive 70 kilometres?

2 Tia now finds out that it is more accurate to say that 1 mile is approximately equal to 1·61 km. Can you work out if this makes a difference to:

   a how long a marathon route is in kilometres? (look back to **question 1**)

   b how far Tia's mum can run in an hour at 9 minutes per mile?

3 Molly is training for a cycle race, so she is trying to do longer and longer rides each time she trains. These are the courses she can follow:

   5 km   3 miles   4500 m   4 miles   3·7 miles   4·49 km

In which order should Molly ride these courses?

Converting units of mass and volume

Practice

1 Ravi is preparing drinks for a party. Each glass holds 0·4 ℓ. Can you work out how many full glasses he can get from a full 1 imperial gallon jug?

I only want full glasses.

I gallon/8 pints — 4·5 ℓ

7 —
— 4
6 —
5 — — 3
4 —
3 — — 2
2 —
— 1ℓ
1 pint —

2 Tia, Ben, Ravi and Molly are going on a plane. They have each packed a suitcase and weighed it, but their scales only show masses in pounds (lbs).

Name	Mass of suitcase (lbs)
Ben	42
Molly	33
Ravi	44
Tia	55

I know that 1 lb is roughly 450 g.

a What is the approximate mass of each suitcase in kilograms?

b If the maximum permitted weight per case is 20 kg, what would you advise the children to do to ensure that they can all bring their cases?

Going deeper

1 In the UK fuel costs £1·15 per litre. In the USA fuel costs $2 per US gallon and a US gallon is approximately 3·8 litres. £1 is approximately $1·25. How much less will it cost to buy 38 litres of fuel in the USA than the UK?

2 A beekeeper used to sell honey in $\frac{1}{2}$ lb jars for £1. He now sells in metric quantities, in jars holding 250 g. Do you get more or less honey for £1 in a 250 g jar than a $\frac{1}{2}$ lb jar? Tip: look at what Molly says above.

Properties of number

Practice

1 Ben and Tia are playing 'Guess the multiple'. Tia's number is a multiple of 2, 3, 4, 5 and 6. Can you work out which number she was thinking of?

Is it a multiple of 2?

2 Ben chose the number 42 in the game. How many questions would you need to ask to work out that this was the number he was thinking of? Is that the smallest number of questions you would need?

3 What is the smallest number of questions you would need to work out that Tia was thinking of the following numbers?

a 63   b 78   c 35   d 96

e 33   f 28   g 81

4 Can you work out and write down all the factors of 90?

Going deeper

1 What do you think is the best strategy for playing the 'Guess the multiple' game in **question 1**? Can you explain why?

2 Kyle says, "There must be three times as many multiples of 6 as there are of 2." Do you agree? Can you explain why?

3 Lia draws a picture of a group of spotted dogs. Some of the dogs have 3 spots and the others have 7 spots. If the group of dogs have 53 spots altogether, how many dogs are in the group?

Lowest common multiples

Practice

1 Ben, Molly and Ravi started saving at different times, but now they have all saved the same amount. What is the smallest amount of money they could each have saved to make this true?

2 a Can you choose two numbers between 2 and 20, and find their lowest common multiple (LCM)?

 b Can you now list all their common multiples between 50 and 150?

3 Which pairs of numbers between 2 and 20 will have a LCM that is also less than 20?

Going deeper

1 How could you work out the LCM of 2, 3 and 5, just using number rods?

2 What is your best method of working out the LCM of 2, 3 and 5 **without** using number rods?

3 Which pair of numbers between 2 and 20 will have the most common multiples between 50 and 150? Can you explain why?

4 Can you explain how this work on LCMs might help you compare and order fractions that have different denominators?

Highest common factors

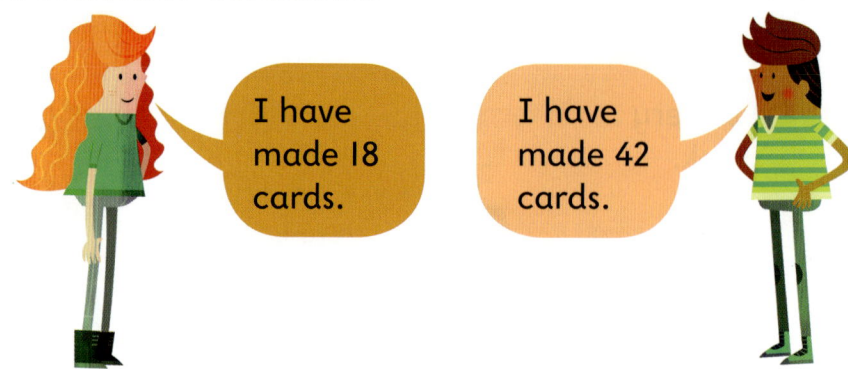

I have made 18 cards.

I have made 42 cards.

Practice

1 **a** Molly and Ravi want to sell all of their cards in packs with the same number of cards in, without mixing their cards. How many could they put in each pack?

   b What is the biggest pack they could make?

2 **a** Can you choose two numbers between 20 and 100, and find their highest common factor (HCF)?

   b Can you write down all their common factors?

3 Can you find five number pairs between 50 and 100 that have different HCFs less than 20?

4 Can you find the HCF of 34, 119, 68 and 85?

Going deeper

1 How could you work out the HCF of 18 and 27, just using number rods?

2 What is your best method of working out the HCF of 6, 15 and 24 **without** using number rods?

3 Anika's school has decided to tile the floor of the new library, which is 12 m long and 8 m wide. What are the biggest square tiles that could be used?

4 How might working out HCFs help you to simplify fractions? Can you explain this?

Prime and composite numbers

Practice

1 The number rods above show fraction walls for 11 and for 12. What can you say about the differences between these two walls?

2 Which other numbers less than 25 will have fraction walls like that for 11? Can you explain?

3 Which other numbers less than 25 will have fraction walls like that for 12? Can you explain?

4 a What are numbers like 11 called?

  b What are numbers like 12 called?

Going deeper

1 How many different rectangular arrays can you make with 12 counters?

2 Now try arranging 11 counters into rectangular arrays. How many can you make?

3 How many rectangular arrays do you think you would be able to make:

  a with 19 counters

  b with 21 counters?

Can you explain this?

4 Stephen says that he can make any composite number from a collection of prime numbers: for example 8 is four 2s, 14 is two 7s. What do you think? Explore some composite numbers and see if this is always true.

Exploring multiplying

Practice

1 Look at this sign. How many apples are there altogether?

2 If you know how to calculate 16 × 8 mentally, which other calculations could you do in your head?

3 Can you work out how many of each fruit are in each of these crates?

16 bags in each crate
8 apples in a bag

a 15 bags 5 in a bag

b 17 bags 6 in a bag

c 18 bags 7 in a bag

Going deeper

1 How could you work out 19 × a number? What number facts could you use? Is there another way? Share your method with your partner.

2 How could you work out 8 × 17? What if you cannot remember 8 × 7?

How can you use your number facts to work it out?

3 How many different ways can you solve 9 × 17? You can draw grids to support your thinking.

Multiplying decimals

Molly and Ravi are discussing how far they each walk to school and back.

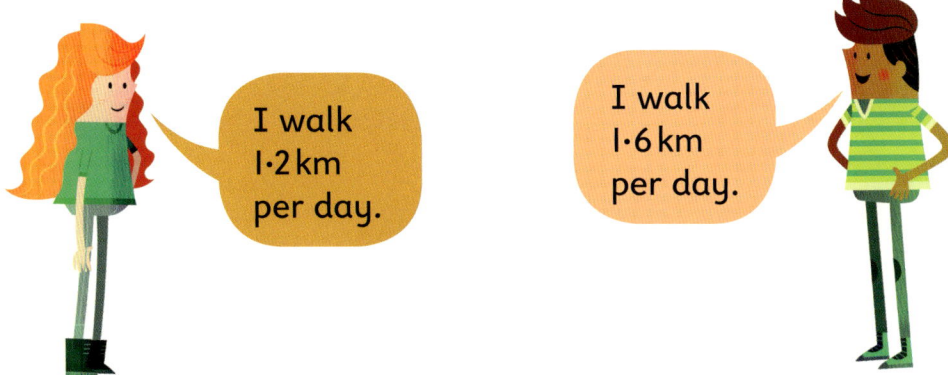

I walk 1·2 km per day.

I walk 1·6 km per day.

Practice

1 a How far do Molly and Ravi each walk to school and back in a week? Can you explain how you worked this out?

Remember there are only 5 days in our school week.

   b Can you work out how far Molly and Ravi each walk to school and back in two weeks?

2 Can you work out how far Lauren, Tamal and Paul each walk to school and back in a week?

Lauren	1·5 km per day
Tamal	0·9 km per day
Paul	2·3 km per day

Going deeper

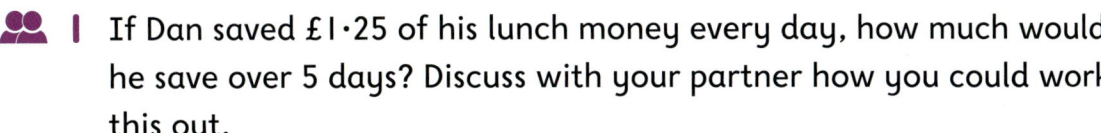

1 If Dan saved £1·25 of his lunch money every day, how much would he save over 5 days? Discuss with your partner how you could work this out.

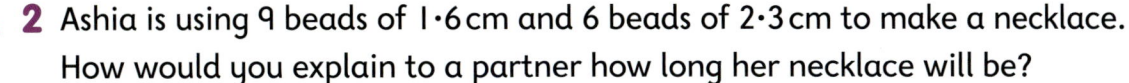

2 Ashia is using 9 beads of 1·6 cm and 6 beads of 2·3 cm to make a necklace. How would you explain to a partner how long her necklace will be?

Exploring dividing

Year 5 children are all at a theme park. There are 118 children on the trip.

All the children want to go on the Tower Ride but it only takes 9 children at a time.

Tower Ride

Practice

1 How many times will the ride need to run so that all the children have one turn each? Think about whether you need to round up or down.

 2 Can you sort the numbers in the table into those which are multiples of 5, 6, 7, 8 or 9? Use your mental dividing methods to help you and record your calculations with a partner.

130	153	190	115	165	125	138
176	171	110	133	152	154	156
155	102	182	195	114	196	117
136	186	203	147	184	174	185
119	170	128	104	161	132	145

Going deeper

1 Look at **question 2**. How can you check these dividing calculations?

2 Choose a number from the table in **question 2**. Can you sensibly partition the number and divide each part to make it simpler?

3 Can you make up some dividing calculations for your partner to solve using a mental method?

Dividing further

We have £7·45 to share equally between the three of us.

Practice

1 a Can you work out how much each child will get?
    How much will be left over?

   b What other ways can you find to divide £7·45 by 3?

2 Can you work out £9·25 divided by 5?

 3 Discuss different methods for working out £12·75 divided by 7.

Going deeper

 1 Finn is wrapping 6 identical presents, but he only has one piece of ribbon, which is 3·66m long. If he cuts the ribbon into 6 equal pieces, how long will each of the 6 pieces be? Discuss how you will work this out.

Can you find another way of doing it?

2 A jug of squash holds 1·8 litres. Can you work out how much squash would be in each cup if you shared the amount equally between 6, 8, 9, 10 and then 12 cups?

Dividing with remainders

Practice

Each table of children gets a pack of 18 mini chocolate bars.

1 On Tia's table there are 4 children.

    a How many whole bars will each child get?

    b Can you say how they could share the remainder equally?

 2 On another table there are 5 children. Discuss how they could share out their pack of 18 mini bars exactly.

Going deeper

1 If 5 children shared mini chocolate bars equally, and each child received $2\frac{2}{5}$ bars, how many bars do you think they started with altogether? Can you explain?

2 If there were 20 apricots, how many would each child get if they were shared between:

    a 6 children   b 8 children   c 12 children?

 3 Play this game with a partner.

    • Write down all the numbers between 1 and 30. Each number can only be used once.

    • Each choose two numbers and divide the larger by the smaller. Write your answer as a whole or mixed number.

    • Do this five times each. The winner each time is the person who produces the biggest number in their answer.

Short dividing

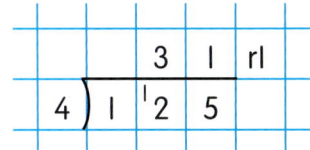

Practice

 1 Can you explain how to turn a remainder into a fraction when short dividing? You can use the example above to explain.

2 How would you solve 318 ÷ 5 using short dividing and turning the remainder into a fraction?

 3 Choose a number from List A and a number from List B to divide it by. Write down the calculation.

> **A:** 176, 215, 325, 450, 592

> **B:** 4, 5, 6, 8, 9

   a Can you predict whether the calculation will divide exactly or give a remainder? Explain your thinking.

   b Now solve your calculations, turning any remainder into a fraction.

   c Do this five times.

4 Can you express any of your remainders in **question 3** as a decimal?

Going deeper

 1 Using the numbers in **question 3**, how could you find the biggest possible answer and the smallest possible answer? Test out your ideas.

2 a Can you make up some dividing calculations that will have $\frac{2}{5}$ in the answer?

   b Can you make up some other calculations that will have $\frac{1}{6}$ in the answer?

Linking improper fractions to dividing

Practice

1 If one muffin is $\frac{1}{12}$ of a tray, what proportion of a tray would 5 muffins be? Can you write this as a fraction?

2 If one muffin is $\frac{1}{12}$ of a tray, what proportion of a tray would these be?

   a 11 muffins   b 17 muffins

   c 24 muffins   d 36 muffins

3 Can you work out how many egg boxes of 6 you could fill if you had 42 eggs? How could you illustrate this?

4 Can you divide 48 yoghurts exactly into packs of 8? How many wholes are in 48 eighths?

5 Can you say what $\frac{24}{6}$ and $\frac{24}{4}$ are as whole numbers? Can you explain your answers using a drawing or representation?

....

Going deeper

1 Can you give an improper fraction that is equivalent to:

   a 5   b 6   c 7   d 8?

2 Can you find three different ways to write the number 12 using improper fractions? Can you illustrate one of these ways on a number line?

3 If you had 50 marbles and they were put into containers that held 12 marbles, how many containers would you need?

Changing improper fractions to mixed numbers

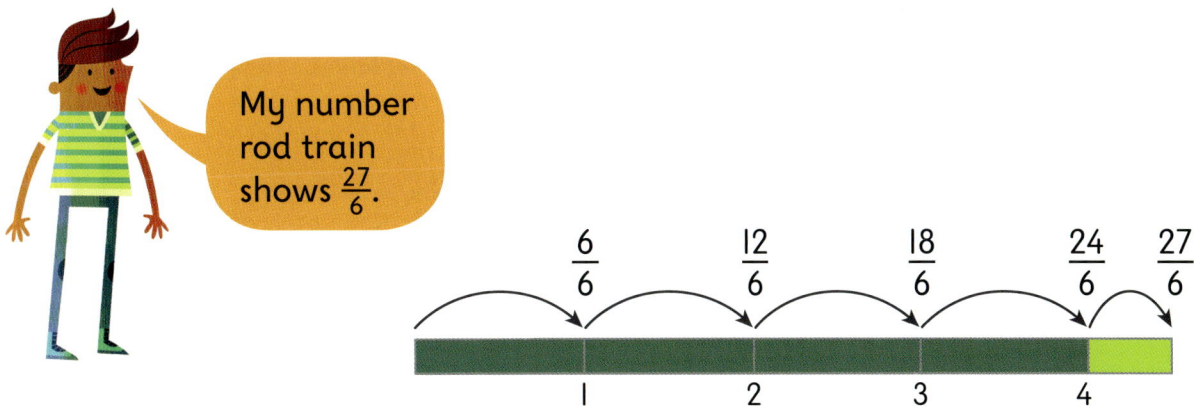

My number rod train shows $\frac{27}{6}$.

Practice

1 Can you write $\frac{27}{6}$ as a mixed number?

2 a Can you make a number rod train to illustrate $\frac{45}{8}$ as a mixed number?

   b How could you change $\frac{45}{8}$ to a mixed number in your head? Can you explain to your partner?

3 Can you convert these improper fractions to mixed numbers? Explain to one another how you are working them out.

   a $\frac{59}{6}$ b $\frac{89}{7}$ c $\frac{103}{10}$ d $\frac{128}{5}$

4 Can you work out some improper fractions that are equivalent to:

   a $3\frac{1}{2}$ b $5\frac{1}{6}$ c $2\frac{2}{3}$?

Going deeper

1 a Choose two numbers from the list below to make an improper fraction. What is the largest improper fraction you can make?

   28, 32, 15, 12, 3, 8, 5, 9

   b What is the smallest improper fraction you can make with the numbers listed above?

   c Can you make a list of improper fractions using the numbers above, then order them from smallest to largest?

Understanding angles

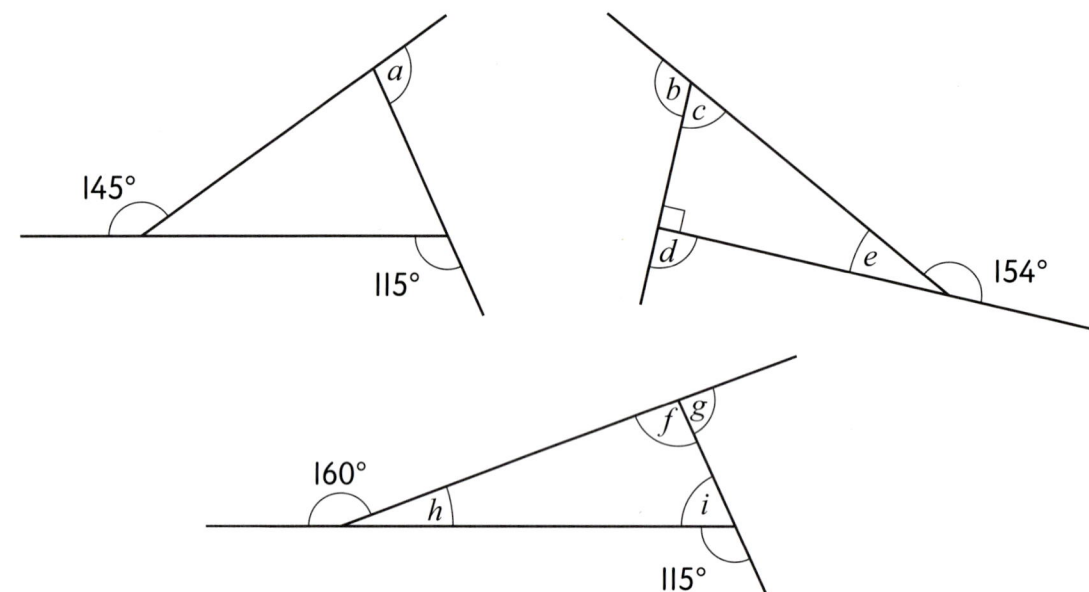

Practice

1 Can you say which of the angles above are interior angles and which are exterior angles? Can you talk about how the angles are related?

2 Find the size of angles *a–i* on the diagrams above (not to scale). For each diagram, can you say which order you worked the angles out in and why?

Going deeper

1 The three interior angles and the three exterior angles of a mystery triangle are mixed up.

150°   48°   30°

132°   102°   78°

a How can you work out which angles are which? Discuss what you think the steps to solving this problem are with your partner.

b Now, can you put this plan into action and find out which three angles are the interior and which are the exterior angles of the triangle?

c Can you sketch and label the triangle?

Exploring angles in triangles

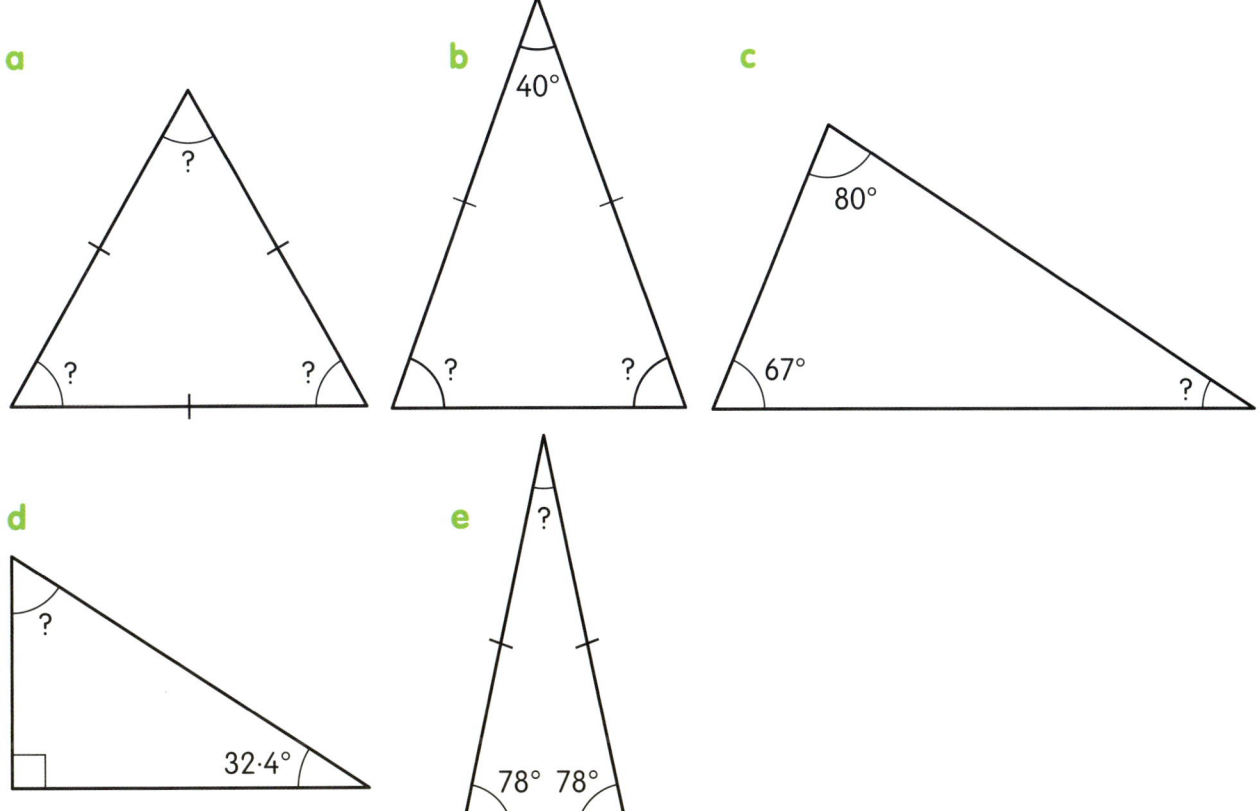

Practice

1 What type of triangle is each of the triangles above? Explain how you know.

2 Can you find the missing angles in the triangles above?

Going deeper

1 Can you sketch an isosceles triangle with an obtuse angle, where all the angles in the triangle are multiples of 15 degrees?

2 a Now can you sketch an isosceles triangle **without** an obtuse angle, where all the angles are multiples of 15 degrees?

   b What could the angles in your triangle be? How many solutions can you find?

Exploring angles further

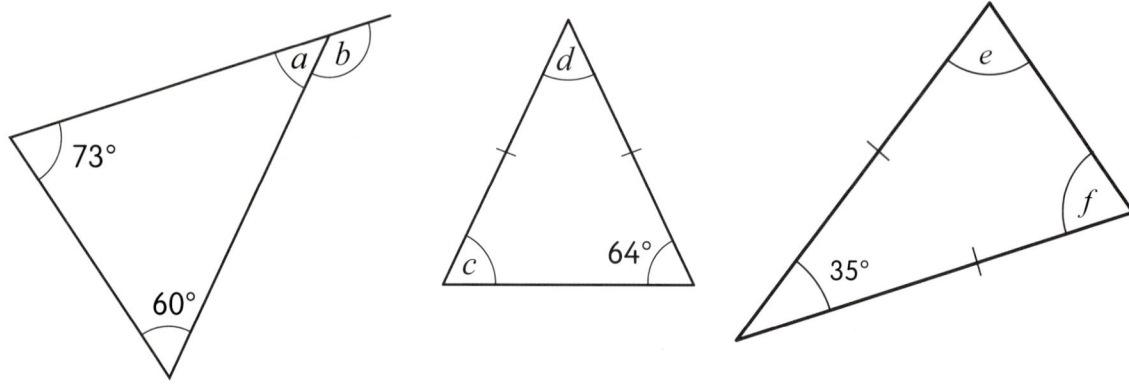

Practice

1 Can you find the missing angles in the triangles above?

2 Can you explain how you worked out the answers to **question 1**?

Going deeper

1 Dipa draws a triangle with interior angles of 60° and 40°. What is the third angle?

2 Can you draw a triangle with angles of 38° and 90°?
  What should the third angle be? Try measuring it using a protractor.
  How accurate were you?

3 Think about the following statements about triangles. Which do you think could be true and which false? Can you explain why?

a My triangle has two right angles.

b My triangle has two obtuse angles.

c The number of degrees in each interior angle of my triangle is odd.

d The number of degrees in each interior angle of my triangle is even.

Exploring angles in quadrilaterals

NOT TO SCALE

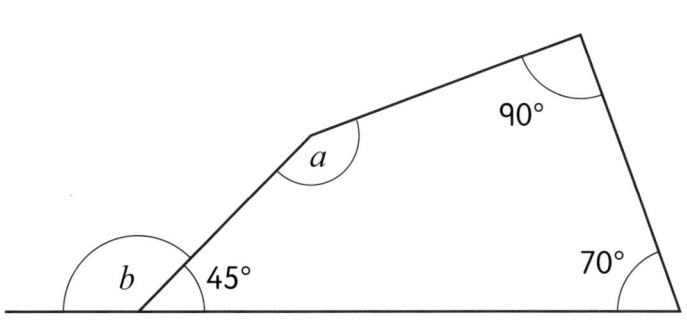

Line of symmetry

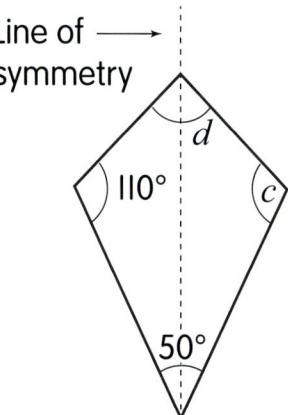

Practice

I Can you find the missing angles in these quadrilaterals?

Going deeper

I Try to find the interior angles of the children's secret quadrilateral from these clues:

Its angles are all multiples of 10 degrees.
The two largest interior angles are the same size.
The smallest angle is one third of the size of the two largest angles.

2 Can you explain how you worked out the answer to **question I**? What rules did you use?

3 How many different quadrilaterals are possible?

Proportion and ratio

Muesli

Serves 8

60 g dried cranberries

200 g oats

40 g brown sugar

80 g toasted almonds

100 g wheat bran

I am making muesli for 16 people.

Practice

1 Can you work out what Ravi needs to make 16 servings of muesli?

2 Tom follows the same muesli recipe and uses 400 g of wheat bran. How much of each of the other ingredients does Tom need to use?

3 Sita uses 500 g of oats to make the same museli. Can you work out how much she needs of each of the other ingredients?

Going deeper

1 How much of each ingredient do you need to make 10 servings of muesli?

2 How many servings could you make with 105 g of dried cranberries, if you had the other ingredients in the correct proportions?

3 If you have 720 g of ingredients altogether, how many servings is this? Can you explain how you know?

4 What proportion of the muesli is made of oats?

Making scale drawings

Practice

1 Sophie wants to make a scale drawing of her garden, which has two rectangular flowerbeds. Can you model or accurately draw the flowerbeds above? Use the scale 2 cm represents 10 cm.

2 Explain how you would draw this flowerbed using the scale 1 cm represents 10 cm.

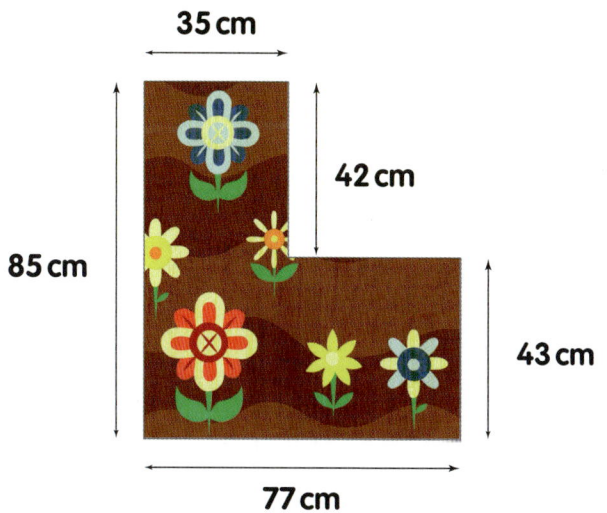

Going deeper

1 Sophie's garden is 12 m wide and 18 m long. Her scale drawing of it is 4 cm by 6 cm. Can you make a different scale drawing of Sophie's garden? Ask your partner to work out the scale you have used.

2 Try drawing a rectangle to different scales. What do you notice about the areas of your rectangles?

Solving problems involving rates

Ice skating	Peak (weekends)	Off-peak (weekdays)
Under 5s (per 30 minutes)	£1	50p
Over 5s (per 30 minutes)	£1·25	80p

I'm 9 years old and today is Saturday.

Practice

1 How much will it cost if Tia stays at the ice rink for 2 hours?

2 How much will it cost if she skates for $3\frac{1}{2}$ hours on Monday?

3 Can you work out how much it costs for Tia's 4-year-old cousin to skate for $1\frac{1}{2}$ hours at the weekend?

4 On Wednesday Tia and her cousin skate from 3:45 p.m. until 5:15 p.m. Can you work out how much it costs them?

Going deeper

1 It costs £12·50 for 2 children to skate at the weekend. Can you work out how long they stayed and whether they were under or over 5 years old?

2 Can you work out five different combinations for spending up to £10? For example, 3 children over 5 years old skate for 30 minutes off-peak: 3 × 80p = £2·40.

3 Jake is 10 years old. He needs to hire ice skates and skates for $2\frac{1}{2}$ hours on a weekday. How much does it cost altogether?

Skate hire £1·75

Solving conversion problems

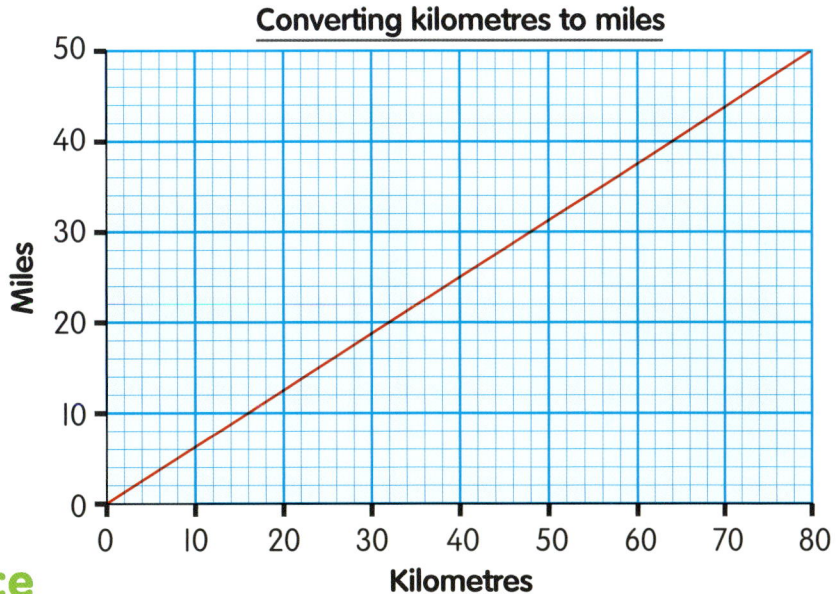

Converting kilometres to miles

Practice

1 Explain how you would use the graph to find out how many kilometres there are in 10 miles.

2 Can you use the graph to work out how many miles there are in 48 km?

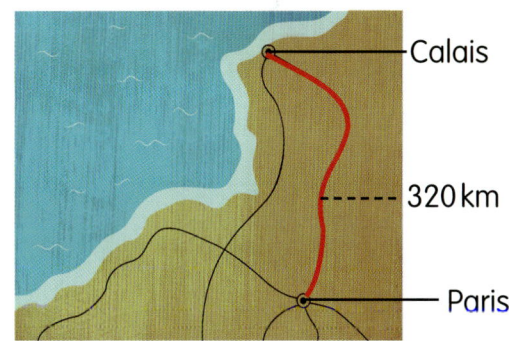

3 Choose a distance that appears on the graph's scale and challenge your partner to convert it to miles or kilometres.

4 Can you calculate how many **miles** it is from Calais to Paris?

Going deeper

1 The distance from Calais to Lille is 68 miles. Can you work out roughly how many **kilometres** this is and explain how you worked it out?

2 Which is further: Calais to Nantes (600 km) or Calais to Dijon (350 miles)?

3 Carl's family drives 75 miles from London to Dover and then 320 km from Calais to Paris. How far did they drive altogether in kilometres?

Understanding 'per cent'

My great-grandma is 100 years old today. 100 years is called a century.

Practice

1 Ben is 10 years old. What percentage of his great-grandma's age is he?

2 Ben's cousin is double Ben's age. What percentage of her great-grandma's age is she?

3 Ben's dad is 50. Can you work out what percentage of their great-grandma's age that is?

Going deeper

1 Ben's mum is less than half his great-grandma's age but greater than 25% of her age. What age range must his mum fall into?

2 Can you give a percentage that is less than $\frac{1}{4}$?

3 Which is greater, $\frac{3}{4}$ or 80%? Explain how you know to a partner.

Different ways to make 50%

There are lots of different ways to show 50%.

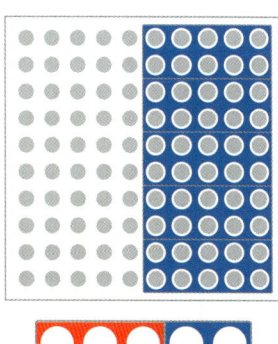

Practice

1 Can you find some other ways to show 50% using Numicon Shapes or number rods?

2 Can you find five different ways of showing 25% using Shapes or rods?

3 Now can you find four different ways to show 40%?

Going deeper

1 If 25% of 80 is 20, how can you find 75%? Explain how you know.

2 20% of a number is 10, so what must 100% of that number be?

3 Can you make up four more questions like the ones above for your partner to solve?

Calculating percentages

Practice

1 How much does the jumper cost in the sale? Can you explain how you worked it out?

2 Can you work out how much the other items cost in the sale?

3 What if the sale was 75% off all full price items? How much would each item cost now? Use this diagram to help you and discuss your methods with a partner.

25%	75%

Going deeper

Original price		New price
£28		?
?	75% Off	£10
?		£14
£35		?

1 Can you copy the table and work out the missing values?

2 A shop offered 20% off all items of clothing. The new price of a T-shirt was £20. What was the original price? Discuss with a partner how to solve this.

3 A bottle of shampoo contains 260 ml, which includes 30% extra free. Discuss how much the original bottle must have contained.

Decimal, fraction and percentage equivalents

I am going to draw a number line to help me order these different amounts.

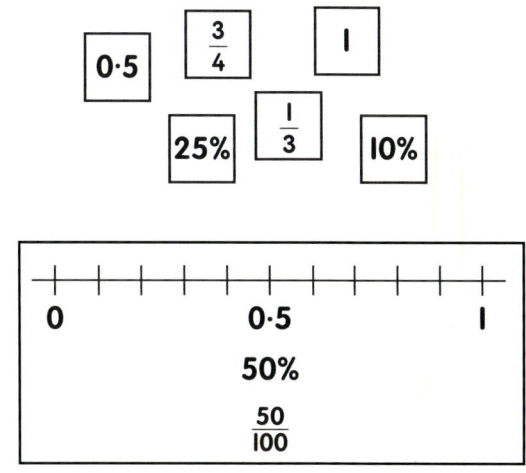

Practice

 1 With a partner, can you make or draw a number line and label it with all the fractions, decimals and percentages on the cards above?

2 Which decimal and fraction are equivalent to 10%?

3 Can you mark where 0·2 would be on your number line?

 Use your $\frac{1}{3}$ mark to help you!

4 Can you mark where 66% should go (approximately)?

Going deeper

1 How many different ways can you express these percentages?

a 75%   b 40%

2 Can you give a percentage that would lie between 0·3 and 0·4?

3 Can you order the amounts below and position them on a number line?

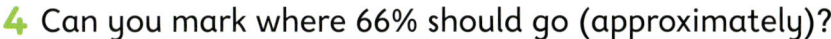

$\frac{4}{5}$ 0·3   80%   60%   0·25 $\frac{1}{10}$

Understanding charts and graphs

I rolled my dice 20 times and got these numbers.

Practice

1 Copy and complete this table using the dice numbers 1–6 above.

Number rolled	Tally	Number of times	Score
1	II	2	2
2			

2 Can you estimate the total score for all Tia's throws? Now work it out. How close was your estimate?

3 Are Tia's results what you would expect? How many times would you expect to roll each number on the dice?

4 Roll a dice 20 times and record your scores in a table like the one above. Were your results what you expected?

Going deeper

1 Mike, Gita and Andrew play darts. There are three rounds and the highest total wins. Can you copy and complete the table to show that Andrew came second? How did you work out the missing numbers?

Name	Round 1	Round 2	Round 3	Total
Mike	17	18	12	
Gita	15	17		45
Andrew		9	17	

Using line graphs

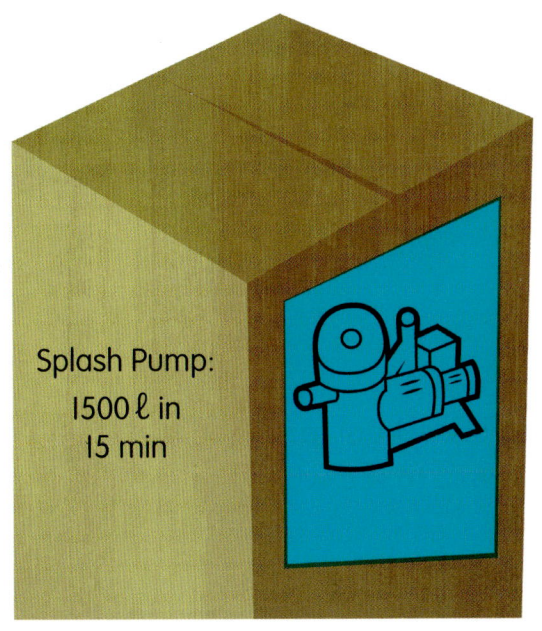

Splash Pump:
1500 ℓ in
15 min

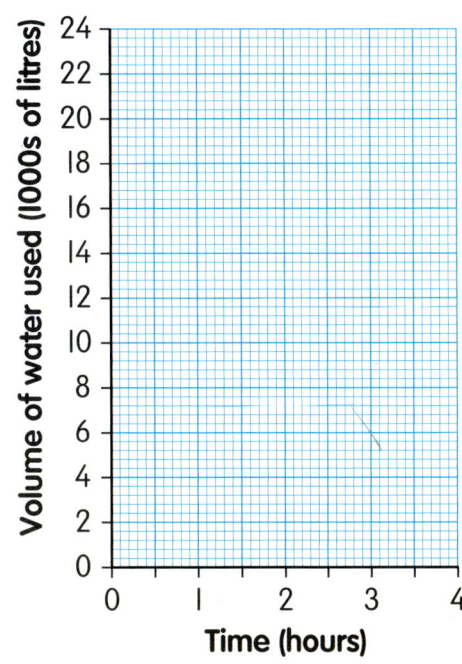

Practice

1   Pete's Pumps provide equipment for swimming pools. Draw a line graph
    on graph paper, matching the scale above, to show
    how much water the Splash Pump can pump in 4 hours.

2   Pete's Pumps also provide an Eco-Pump. Add a second
    line to your graph in a different colour and label it to
    show the flow of the Eco-Pump.

3   Can you use your graph to find out how long it would
    take to fill a 12 000 ℓ pool using:

    a the Splash Pump   b the Eco-Pump?

Eco-Pump:
4000 ℓ
per hour

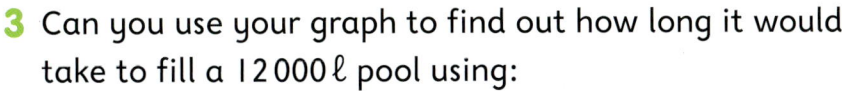

Going deeper

1   Natalie is filling her 15 000 ℓ swimming pool. How much longer will it
    take if she chooses the Eco-Pump instead of the Splash Pump?

2   Jon fills his pool and finds it was 30 minutes faster using the Splash
    Pump instead of the Eco-Pump. What is the capacity of Jon's pool?
    How do you know?

Measuring temperature over time

We bought ice cream from the supermarket at ⁻28°C.

We need to serve it between ⁻14°C and ⁻12°C.

Practice

Ben and Molly are running an ice cream stall at the school fair. They record the temperature of the cool box as it warms, to see when the ice cream is ready to serve.

Time	09:00	09:15	09:30	09:45	10:00	10:15	10:30	10:45
Temperature (°C)	⁻28	⁻24	⁻20·5	⁻18	⁻16	⁻14·5	⁻13·5	⁻13

1 Can you show this data on a graph? Use graph paper and label time on the x-axis and temperature on the y-axis.

2 Using your graph, can you suggest when the ice cream stall should open for business?

3 Can you explain to your partner how long Molly and Ben had to wait to serve the ice cream?

4 What was the change in the ice cream's temperature in this time?

Going deeper

1 If the temperature of the cool box continues to change in a similar way, when should they stop selling the ice cream? Can you explain why you think this?

2 At another school fair, Ben and Molly are asked if they could open their stall at 09:00.

   What time do you think they should buy the ice cream? Can you explain why?

Presenting and comparing data charts

	100 m	400 m	Long jump	Tennis ball throw
Ethan	14 s	68 s	2 m	10 m
Mia	14·5 s	67 s	2 m 30 cm	10 m 40 cm
Lizzie	15·5 s	82 s	2 m 40 cm	9 m 60 cm
Ahmed	17 s	80 s	1 m 95 cm	12 m

These are the results of our sports day!

Practice

 1 Using appropriate scales, draw four bar charts to represent the information in the table above.

2 What do the bar charts show about the athletes?

3 Who do you think is the best athlete? Do you and your partner agree?

Going deeper

1 In an event, 24 prizes are hidden around the playground and the children have to find as many of them as they can. The game ends when all the prizes are found. Draw a pie chart and a bar chart to show the results of the game.

Tia — I found 6.

Ben — I found 4.

Ravi — I found 8.

Molly — I found all the rest!

 2 Which chart do you think shows the data most helpfully? Can you explain why to your partner?

Solving problems with fractions, decimals and percentages

Practice

1 Can you write 30% as a fraction in its simplest form? Now can you write it as a decimal?

 2 Take turns to cover part of a baseboard with some Numicon Shapes.

   a Ask your partner to say what percentage of the baseboard you have covered.

   b Can you also write these percentages as both fractions and decimals?

3 Now cover part of a baseboard with some Numicon Shapes again.

   a Ask your partner to say what percentage is **not** covered.

   b Can you write these percentages as both fractions and decimals?

Going deeper

1 Can you put the following fractions, decimals and percentages in order of size, beginning with the smallest?

$$\frac{2}{5} \quad 35\% \quad 0·42 \quad \frac{3}{10} \quad 40\% \quad 0·38$$

2 a Where did you last see or hear percentages being used outside school? What were these percentages being used for?

   b Do you think it would have been more useful or less useful to use fractions or decimals instead? Can you explain why, or why not?

Percentages of amounts

Practice

Book signing offer
15% off marked price

1 Can you work out what the price of the book will be after 15% has been taken off?

2 Take it in turns to choose a percentage from List A and a mass from List B and work out the percentage of the mass you chose. Explain your working to your partner and check that you agree with the results.

A: 80%, 20%, 5%, 50%, 8%, 10%

B: 36 kg, 200 g, 72 kg, 3200 kg, 10 kg, 450 g

3 Can you work out how to check your calculations in other ways?

Going deeper

1 What do you think is the best way to work out 75% of anything? Can you explain why?

2 Can you decide the best way to work out 65% of anything? Talk about your decision.

3 Which is heavier, 20% of 29 kg or 0·3 × 25 kg? Try to explain your thinking to your partner.

Equivalents

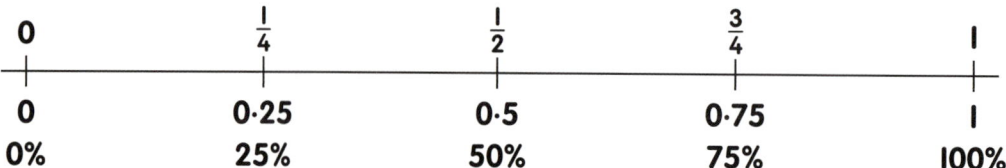

Practice

1 Try drawing a number line like the one above. Can you now mark on it 20%, 40% and 60%, and show these percentages as fractions and decimals as well?

2 Can you work out 50%, 25% and 75% of the following prices?

   a £58   b £4·64   c 96p

   d 8p   e £300   f £48

 3 What do you think is the best method for working out the percentages in **question 2**? Can you explain it to a partner?

4 Can you think of a good way to work out 17·5% of anything?

Going deeper

1 Using two different number rods, how many different ways can you find to illustrate the percentages below? How do you know that you have found all the ways?

   a 25%

   b 20%

2 Jay spent $\frac{1}{4}$ of his pocket money, and Holly spent 20% of hers. Which of them spent the most, do you think?

   Can you say why you think this?

Percentages as proportions, and as operators

Practice

1 Can you draw your own diagrams, like the one on the right, to show as many percentages as possible of:

a 23 ℓ

b 36 m

c 80 kg

d £4

e 72 minutes?

2 Copy the diagram on the right. Can you fill in all the missing distances?

3 Can you explain a quick way of working out the whole distance, just from knowing that $2\frac{1}{2}\% = 5$ km?

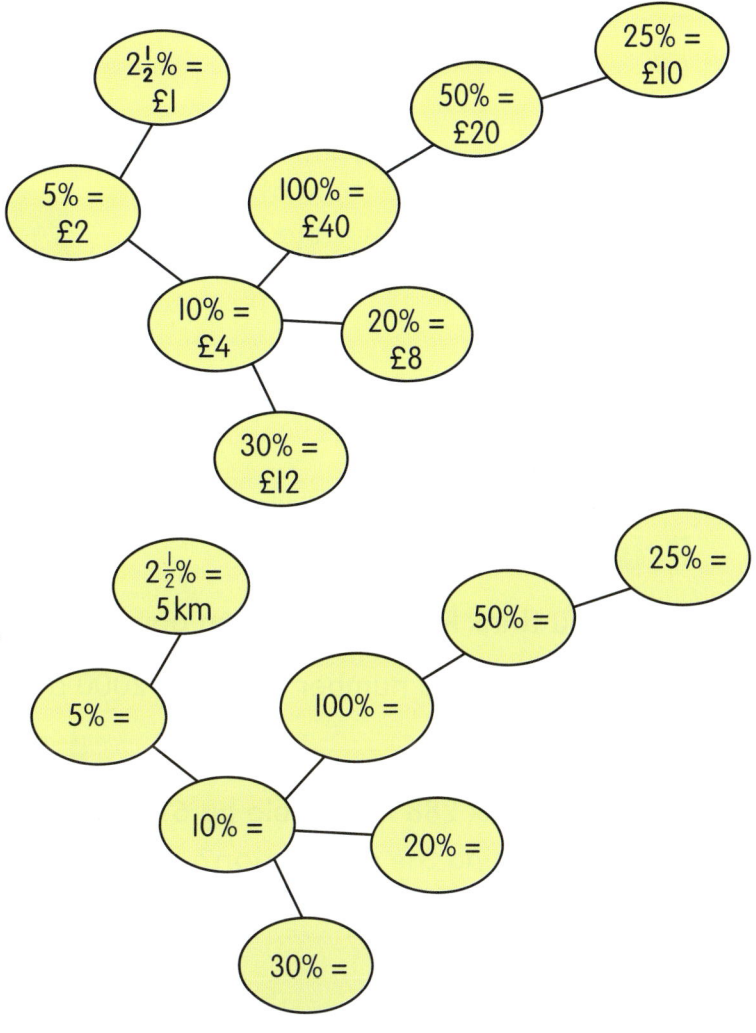

Going deeper

1 What proportion of the positive, whole numbers up to (and including) 100 are:

a even   **b** odd   **c** prime   **d** composite

e multiples of 6   **f** odd multiples of 5?

2 Holly thinks that 29% and 69% are the hardest percentages to calculate of anything. Do you agree? Can you say why, or why not, using examples to explain your thinking?

Looking for patterns and generalizing

I think 312 is divisible by 4.

Practice

1 Do you think Ravi is right? Can you explain why you think so?

2 Can you write a number between 2000 and 3000 that is divisible by:

a 2   b 3   c 4   d 5   e 10?

3 Do you think 288 is divisible by 6? Can you explain why or why not **without** doing the calculation?

Going deeper

 1 Can you show and explain why numbers that are divisible by 5 always end in a '0' or a '5'? You can use Numicon Shapes to show this.

2 Can you explain the rule that all numbers are divisible by 4 if their last two digits are divisible by 4?

3 Can you explain why multiples of 3 are sometimes odd and sometimes even?

Sequences of rod patterns

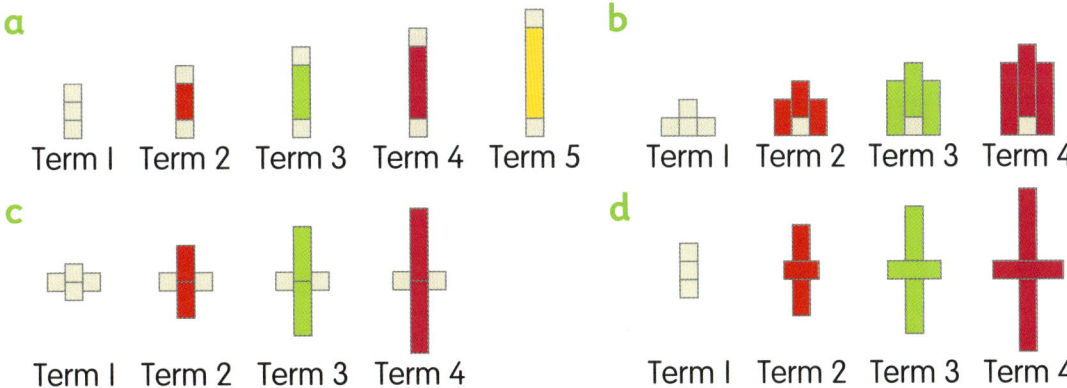

a Term 1 Term 2 Term 3 Term 4 Term 5

b Term 1 Term 2 Term 3 Term 4

c Term 1 Term 2 Term 3 Term 4

d Term 1 Term 2 Term 3 Term 4

Practice

1 Can you work out what the total length of the rods of Term 10 would be in each of the rod sequences above?

2 Can you describe the rod designs that are growing in each of the rod sequences above?

3 Can you write the next term for each of these rod sequences with numbers? For example, Term 6 of sequence **a** will be '6 + 2'.

Going deeper

1 Can you work out what the total rod length of this new sequence would be for Term 10?

2 **a** Can you describe how the hole in the middle of the rod design in **question 1** is growing?

 b How big do you think the hole will be in Term 10?

3 Can you describe any connections you see between the rod lengths and the area of the hole in the rod design sequence in **question 1**?

4 Can you find a new way to calculate the total rod length of Term 20 in the **question 1** sequence?

5 Can you find a way to calculate the total rod length of any term in the **question 1** sequence?

Square numbers

 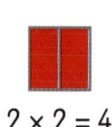   

$1 \times 1 = 1$ $2 \times 2 = 4$ $3 \times 3 = 9$ $4 \times 4 = 16$ $5 \times 5 = 25$

Practice

1 a Using a set of number rods, can you describe the length of each rod in relation to its width? For example, the length of the yellow rod is five times its width.

   b What do you notice? Can you explain?

2 If there were 12 different number rods, how many times longer would the length of the 12-rod be than its width?

3 a Can you use these rod length and width comparisons to explain why four dark pink rods put together side by side make a square?

   b Can you explain why five yellow rods together side by side make a square?

4 Can you find four different ways to write 5^2?

5 Can you find two square numbers that add up to 45 together?

6 Can you find three square numbers that add up to 45 together?

···

Going deeper

1 Can you explain why adding consecutive odd numbers, starting at 1, gives you square numbers as a result?

2 Can you explain why a square number always has an odd number of factors?

1
$1 + 3 = 4$
$1 + 3 + 5 = 9$
$1 + 3 + 5 + 7 = 16$
$1 + 3 + 5 + 7 + 9 = 25$

Cube numbers

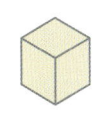
$1 \times 1 \times 1 = 1^3$

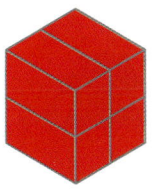
$2 \times 2 \times 2 = 2^3$

$3 \times 3 \times 3 = 3^3$

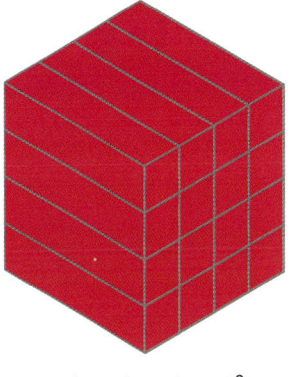
$4 \times 4 \times 4 = 4^3$

Practice

1 Can you write the first seven cube numbers?

2 Can you find two cube numbers that together add up to 407?

3 Can you find three cube numbers that together add up to 755?

4 How many different ways can you find to write 4^3?

Going deeper

1 a Can you make a cube shape using just yellow number rods?

 b What is this cube's total rod length?

2 a Can you make cube shapes using any sized number rods of the same colour?

 b What are their total rod lengths? What do you notice?

3 a Can you connect what you know about square numbers and how you make cube numbers with number rods?

 b Can you explain why these cube shapes give cube numbers?

Area and perimeter

I've found different ways to make a rectangle with these tiles.

Practice

1 Molly has 30 square tiles, each 1 cm long. She makes a rectangle using all the tiles. What perimeter could the rectangle have?

2 If Molly's rectangle was made with 60 squares, what perimeter could it have?

3 Finally, Molly makes a rectangle with an area of 90 squares. How many different perimeters could this rectangle have?

Going deeper

 1 Tia, Ben, Ravi and Molly are planting cabbages in rectangular patterns. How many different arrangements can each child make with their number of cabbages? Explain how you know you've found all the possibilities.

Name	Number of cabbages
Tia	64
Ben	66
Ravi	68
Molly	60

 2 Molly draws a rectangle with an area between 10 and 20 cm², and finds that the perimeter is equal to the area. Can you work out what its dimensions could be? Can you explain the method you used to a partner? Did you answer the question in the same way?

Using area to explore factors, square numbers and prime numbers

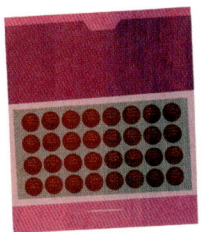

Midi box

64

Maxi box

72

Mega box

100

Practice

1 Choccalot Confectionery produce a mega box containing 100 chocolates. If the chocolates are laid out in a rectangular array, how many different arrays are possible?

2 The maxi box contains only 72 chocolates. Does this have more or fewer possible arrangements than the mega box? Can you list what they are?

3 The midi box contains 64 chocolates. Can you find all the possible rectangular arrays for this number?

4 Given the choice of creating a similar box for 35, 36 or 37 chocolates, which would you **not** choose? Can you explain your reasoning?

Going deeper

1 On squared paper draw a rectangle, length 2 more than its width, and write down its area. Add 1 to this area and note the total.
Repeat this for a range of different-sized rectangles, making sure that the length and width of every rectangle you choose differ by 2.

What do you notice? Can you explain why this happens?

Area, perimeter and decimals

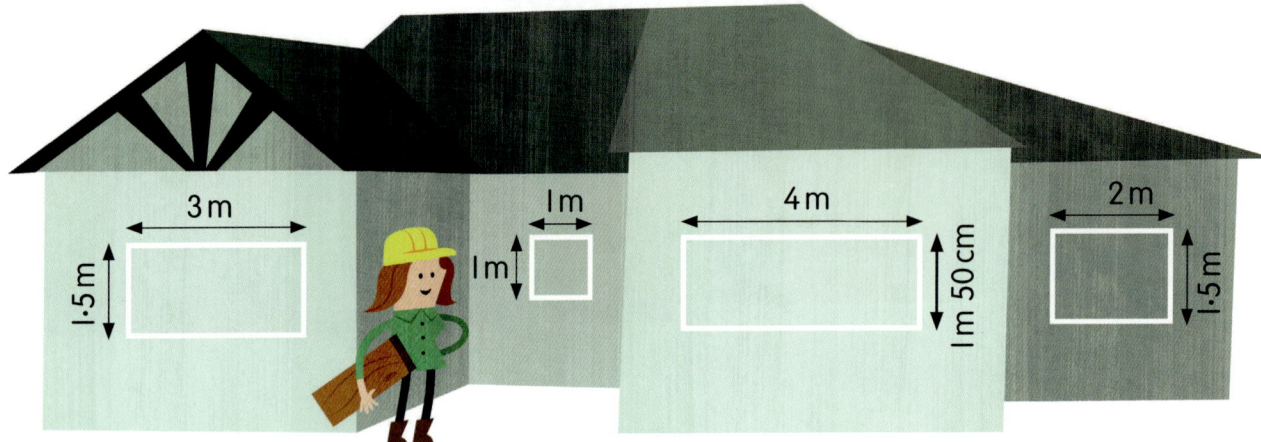

Practice

1 A builder wants to place a board over each of the windows above. How many square metres of board does she need to order in total?

2 Sanjay decides to buy carpet for five rooms in his house. The dimensions of the rooms are given in the table. If carpet costs £20 per square metre, how much will it cost to carpet the rooms altogether?

Study	Dining room	Bedroom	Guest room	Lounge
2m × 2·5m	4·5m × 3m	3·5m × 4m	2·5m × 2·5m	5m × 2·5m

 3 Can you explain to your partner how you worked out your answers to **question 2**?

Going deeper

 1 A rectangle has a perimeter of 26 cm. Its side lengths (in centimetres) are all whole numbers. What could its area be?

How many answers can you find? Can you explain how you know you've found all the possible answers?

2 A rectangle has sides that are all whole numbers between 1 and 10. Its perimeter is 2 less than its area. Can you find its dimensions?

Finding the area and perimeter of composite shapes

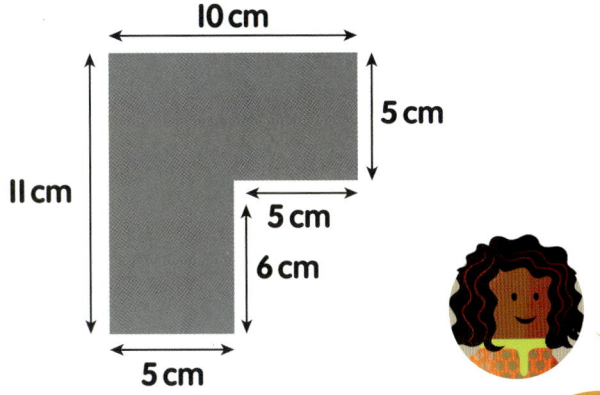

Remember squares are a type of rectangle.

Practice

1 What is the perimeter of the shape above?

2 Can you find two different ways to split this shape into two rectangles and then find the area?

3 Now try to find a way to split the shape into three rectangles. Does this give the same total area? Can you explain why?

Going deeper

1 Al says that he does not need to see all the measurements of the shape above to be able to calculate the perimeter. Just two of them would be enough. Is he right?

2 Draw a 7 cm by 7 cm square. Can you cut a rectangular piece off so that the remaining area is 34 cm^2 and the perimeter is 28 cm?

3 Start with a 7 cm by 7 cm square. How can you cut a rectangular piece off so that the area decreases by 9 cm^2 and the perimeter increases by 6 cm? Explain your ideas to your partner.

Developing written methods of multiplying

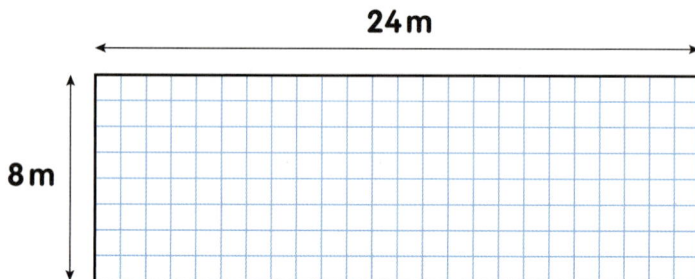
I have a vegetable field that is 24 m long and 8 m wide. Can you help me find the area of my field?

24 m

8 m

Practice

1 Can you copy and divide the rectangle above in some way to help you work out 24 m × 8 m?

2 Now can you find the area of the field using short multiplication?

3 Which of the methods in **questions 1** and **2** did you prefer? Can you use your preferred method to work out the area of each of these fields?

   a Chickens: 18 m × 5 m   b Sheep: 32 m × 6 m   c Cows: 37 m × 9 m

Going deeper

1 Each sheep needs 185 g of feed per day. If there are 8 sheep, how much feed does the farmer need:

   a per day   b per week?

2 Each chicken needs 35 g of feed per day. If there are 6 chickens, how much feed does the farmer need:

   a per day   b per week   c for 4 weeks?

Multiplying decimals using short multiplication

Ravi and Molly are making a square mosaic picture using square stickers.

Each square sticker has a side length of 2·6 cm.

If we place the squares on our picture, we can fit 9 squares in along the top.

Practice

1 Can you work out how wide the finished mosaic will be in centimetres?

2 Ravi makes another square mosaic using 49 square stickers that are each 3·8 cm long. Can you work out how wide his mosaic is?

3 Can you work out these calculations? Estimate first to help you check your answer is sensible.

  a 5·6 × 7   b 3·2 × 6   c 1·9 × 9

Going deeper

 1 Can you estimate and then work out the following? Discuss your different methods with a partner.

  a 0·67 ℓ × 5   b 1·74 pints × 3   c 48·2 kg × 4

2 Can you copy these grids and fill in the missing numbers? Then write the completed multiplying sentences.

a

×	5	?
6	?	3·6

b

×	?	0·4
6	18	?

c

×	2	?	0·05
?	16	2·4	?

Multiplying 2-digit numbers

Practice

1 Ben plants 32 rows of 28 lettuce seeds. Copy this rectangle onto squared paper. Can you divide it up to help you work out the total number of lettuces?

2 Can you and a partner find three different ways to split the rectangle up and then multiply each part to find the overall product?

3 Try to calculate the total number of these vegetables:

a 55 rows of 46 radishes

b 27 rows of 36 leeks.

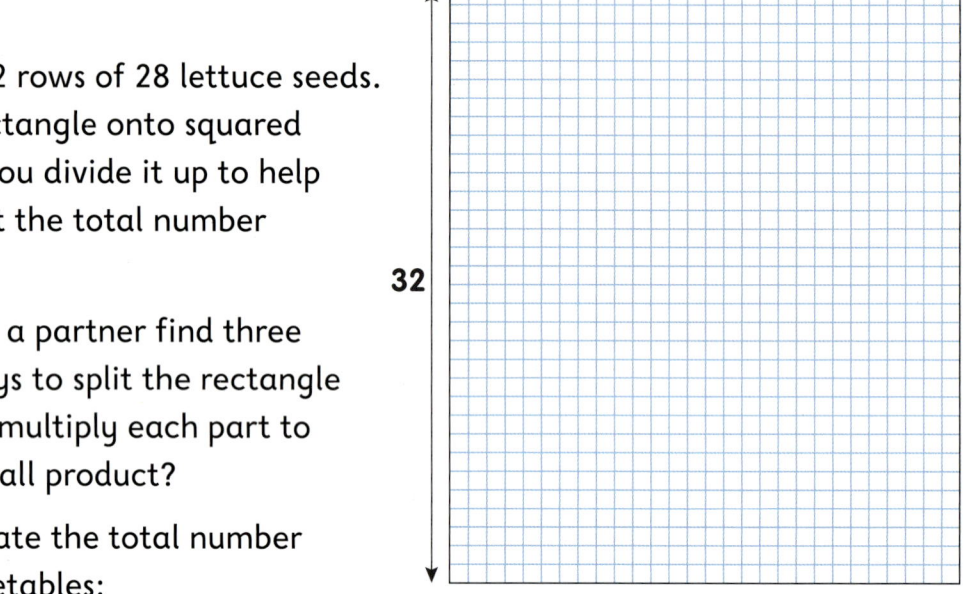

Going deeper

1 Create your own multiplication grids like the one here for a partner, using a calculation of your choice. Then rub out some numbers from the grid. Ask your partner to work out the missing numbers. Take turns.

×	40	?
?	800	60
3	?	9

Using long multiplication

These colouring pencils come in packs of 45.

We have 15 packs. How many pencils are there in total?

×	40	5
10		
5		

Molly works out how many coloured pencils there are using the grid method and Ben solves it using long multiplication.

		4	5
×		1	5
	2	2₂	5
	4	5	0
	6	7	5

Practice

1 Can you copy and complete Molly's grid?

2 Can you explain how the numbers in Molly's grid relate to Ben's long multiplication?

3 Can you work out these calculations? Check your answers using an alternative method.

a 14 × 22   b 27 × 13   c 36 × 28

Going deeper

1 What is 24 × 4·5? Estimate first and then use your chosen method.

2 Which of these calculations would you do mentally and which would you use a written method for? Can you discuss with a partner and then solve them?

a 25 × 12   b 30 × 32   c 46 × 14

d 18 × 20   e 125 × 16   f 5·5 × 22

Volume and capacity

I have eight 4-rods.

Practice

1 Ben arranges all his 4-rods into a cuboid. Can you find the dimensions of all the possible cuboids Ben could make with the 4-rods? What do you notice about the volume of the cuboids?

2 Rhian has twelve 6-rods. What are the dimensions of all the cuboids she can make using all her 6-rods?

3 Can you build a number rod cuboid with a volume of:

a $30\,\text{cm}^3$ b $36\,\text{cm}^3$

c $40\,\text{cm}^3$ d $45\,\text{cm}^3$?

Going deeper

 1 Take turns to choose a number rod colour and an even number from 2 to 12 for your partner. How many different cuboids can they make from that number of rods? What are the volumes of their cuboids?

2 Bethan and Markus challenge each other to create a cuboid with a volume of $60\,\text{cm}^3$ using only one colour of number rod each. Can you work out how many ways they could do this?

Drawing 3D shapes in 2D

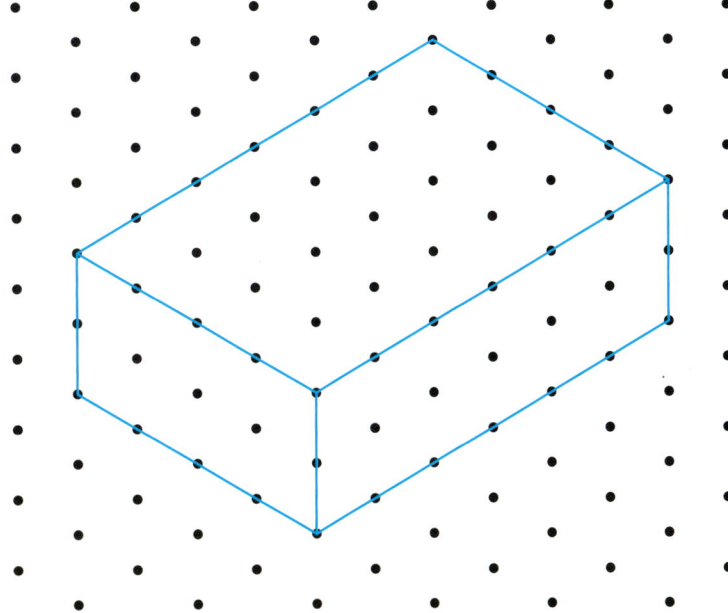

Practice

1 Tia makes a cuboid from 48 cubes and draws it on isometric paper.

a How many other ways can you find of making a cuboid from 48 cubes?

b Can you draw them on isometric paper?

Going deeper

1 Amrit has 45 cubes. He arranges these 1-cm cubes into a cuboid, which is not more than 20 cm long in any direction.

a Can you draw the cuboid he makes on isometric paper?

b Is this the only possibility? Can you explain why, or why not?

2 How many ways can you arrange three cubes into a single shape, if it does not have to be a cuboid? Can you draw your shapes on isometric paper?

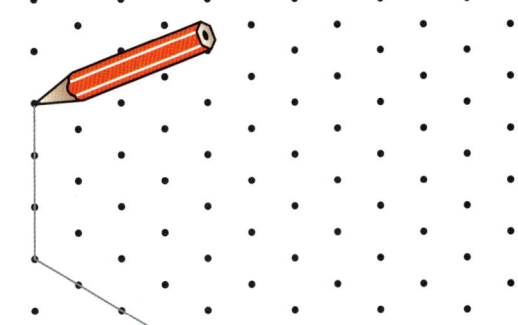

Cubic centimetres and millilitres

Practice

1 How many white 1-rods would you need to add to the cylinder to raise the water level to the 500 ml mark?

2 Can you repeat **question 1** for each different colour of rod up to the 10-rod?

3 Which rod is the odd one out? Can you explain why?

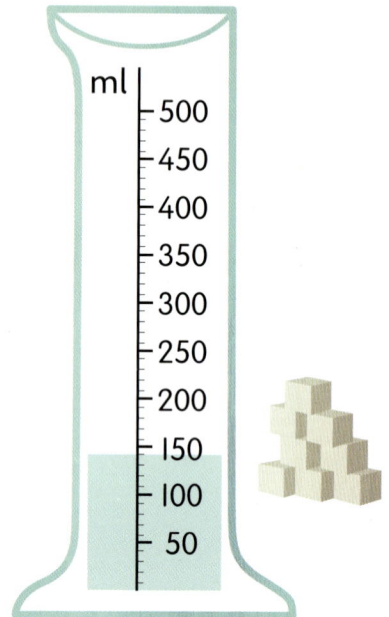

Going deeper

1 Ravi is using 5-rods and 6-rods. He is trying to find out if he can raise the water level in a cylinder from 50 ml to 150 ml by dropping rods in one at a time.

How many different combinations of 5-rods and 6-rods can you find that would do this? (Assume that a 5-rod followed by a 6-rod is the same as a 6-rod followed by a 5-rod.)

2 Ravi next tries to raise the new level further to 179 ml using 6-rods and 7-rods. Is this possible? Can you explain why or why not?

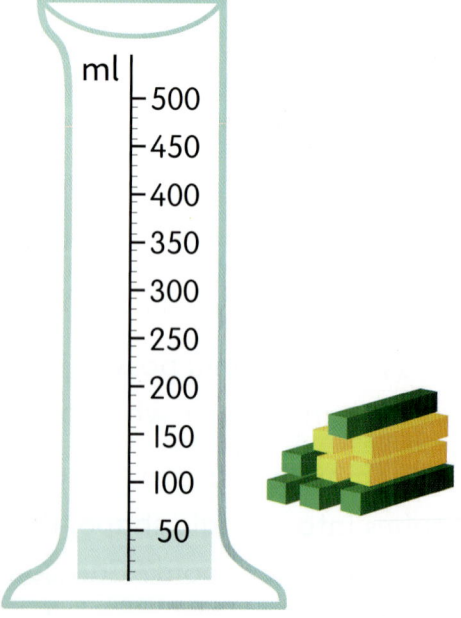

Volume and capacity problems

Practice

1 A rectangular frame on a table has an internal capacity of 120 cm³. Its internal length is 10 cm and internal depth 2 cm. Can you say what the third internal dimension is?

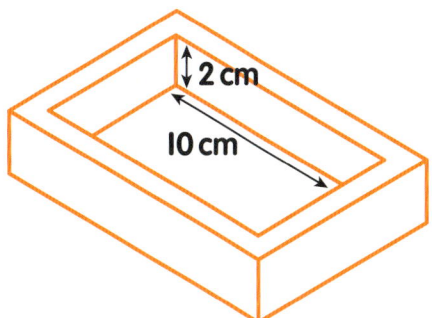

2 The plastic walls of the frame in **question 1** are 1 cm thick. Can you build the frame using number rods?

3 a Can you work out the volume of plastic used to make the frame?

b Can you explain how you worked it out?

Going deeper

1 A recording studio is built in the shape of a cube. To keep out all noise, the walls, the floor and the ceiling are all 1 m thick. The volume of the cube is 216 m³. Can you work out the internal capacity of the studio? Can you draw a diagram to help you?

2 The walls and floor of this large aquarium tank are 1 m thick. Its external dimensions are 8 m by 4 m by 5 m high. What is the internal capacity of the tank in cubic metres?

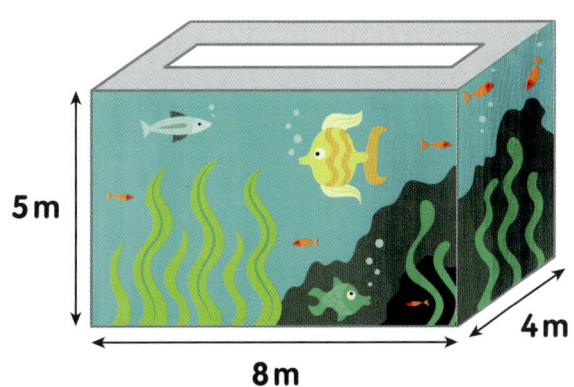

Written methods of dividing

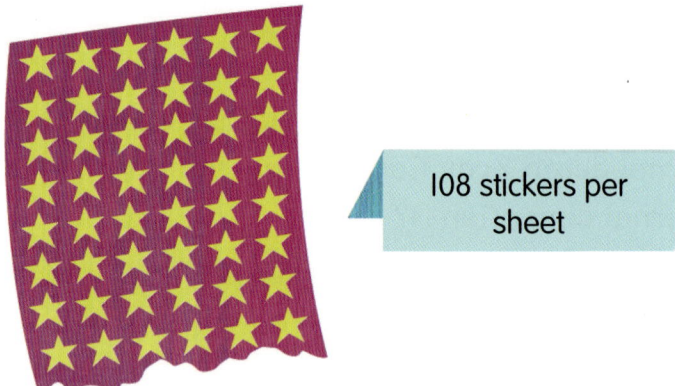

108 stickers per sheet

Practice

1 Can you work out how many **rows** of stickers there must be on a complete sheet?

2 Can you use the short method of dividing to work out how many rows of stickers each of the sheets below will have?

a

126 star stickers

b

192 star stickers

Going deeper

👥 1 Can you think of a scenario where you could use the following calculation? Share your dividing story with a partner.

120 ÷ 15 = ▨

👥 2 Can you make up some other number stories for your partner to solve using the short written method of dividing?

Short written method of dividing

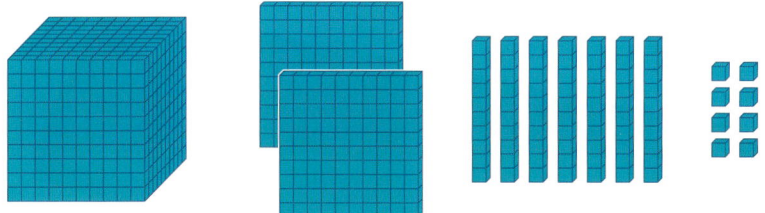

Practice

1 Aisha is using base-ten apparatus to solve 1278 divided by 6. Can you explain how to do this calculation using base-ten apparatus?

2 Can you divide the following numbers by 6? As you do each one, talk through your method with a partner.

a 1290   b 1194   c 1944

d 2560   e 2717

Going deeper

1 George has tried the following short dividing calculation, but has made a mistake.

$$\begin{array}{r} 2\ 7\ \text{r}5 \\ 8\overline{)2\ 1\ 6\ 5} \end{array}$$

Can you explain what George has done and what the correct answer should be?

2 Discuss the two calculations below. Can you see what errors have been made?

a
$$\begin{array}{r} 2\ 3\ 8 \\ 7\overline{)1\ 6\ 7\ 3} \end{array}$$

b
$$\begin{array}{r} 4\ 2\ 3 \\ 3\overline{)^{1}2\ 7\ 9} \end{array}$$

3 Can you work out the correct answers to the dividing calculations in **question 2** above? What calculations could you use to check that your answers are correct?

Dividing with the answer as a decimal

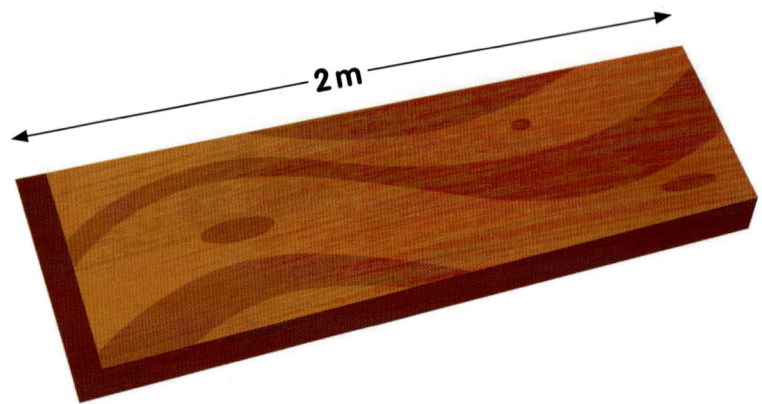

2m

Practice

1 A carpenter has a piece of wood 2 m long and wants to make 3 shelves of identical length. Can you work out how long each shelf will be:

    a to the nearest centimetre   **b** to the nearest millimetre?

2 The carpenter has 3 metres of wood and wants to make 7 shelves of equal length. Can you work out how many metres long each shelf will be, to two decimal places?

3 Can you use short dividing to work out $5\,\ell \div 8$?

Going deeper

1 Which of the following do you think will result in an answer with more than three decimal places? Investigate, and then explain what you find out to your partner.

    a $4 \div 6$ **b** $1 \div 7$ **c** $1 \div 3$

    d $2 \div 8$ **e** $7 \div 9$ **f** $3 \div 5$

2 Can you think of six dividing calculations that you know will result in an answer with decimals?

Dividing money using the short written method

Practice

 1 a Check that if each child on the right pays £6·51, this will cover the full cost of the bill.

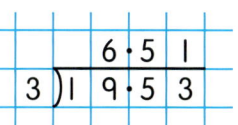

 b The children decide to add a 10% tip of £1·95 to the bill. Can you work out how much each person will pay in total now?

2 Using the information below, can you work out how much each activity costs per person?

a

Swimming
4 children
£19·40

b

Ice Skating
3 children
£16·95

c

Farm Visit
5 children
£47·25

Going deeper

 1 Can you divide each of the following amounts of money by 3, 4 and 5? Explain to a partner which ones you would do mentally and which ones you would use a written method for.

a £18   b £28·80   c £45·60

2 Can you think of some amounts of money that divide exactly by 3, 4 and 5? Give at least two examples of each.

Finding fractions of amounts

Practice

1 a Joe says that $\frac{1}{4}$ of the splodges in the array above are yellow.
      Is he correct? Can you explain how you know?

   b Can you write a dividing sentence that describes the array?

2 Can you use the array above to work out?

   a $\frac{3}{4}$ of 32   b $\frac{3}{8}$ of 32

3 a Can you draw an array to illustrate this fraction fact: $\frac{1}{5}$ of 35 = 7?

   b What other facts can you work out using your array?

Going deeper

1 a Make or draw a 5 by 3 array. What fraction facts about 15 can you
      see in your array? Record these as fraction number sentences, for
      example $\frac{1}{3}$ of 15 is 5. Can you find all the possibilities?

   b Can you make any other arrays for 15? Explain why, or why not?

2 a What different arrays can you make for the number 12? What fraction
      facts about 12 can you write? Can you find all the possibilities?

   b Have you found more fraction facts for 12 than you found for 15?
      Can you explain why?

Finding fractions of quantities

Colour	Fraction of cars made
Red	$\frac{1}{3}$
Blue	$\frac{1}{6}$
Black	$\frac{1}{10}$
Silver	$\frac{2}{5}$

Practice

1 The table above shows what fraction of the cars made in a factory are blue, red, silver or black. If the factory makes 150 cars in a day, how many of each colour are made on that day?

2 How much do you think 3 hours of parking in this car park would cost? Can you explain how you worked this out?

Car Parking
480 spaces
£3 for 4 hours

Going deeper

1 If the factory in **question 1** made 120 blue cars in a day, how many cars would it make altogether on that day? Can you explain how you worked this out?

2 Can you think up some more fraction questions to ask your partner about the car factory in **question 1** or the car park in **question 2**?

Measurement and finding fractions

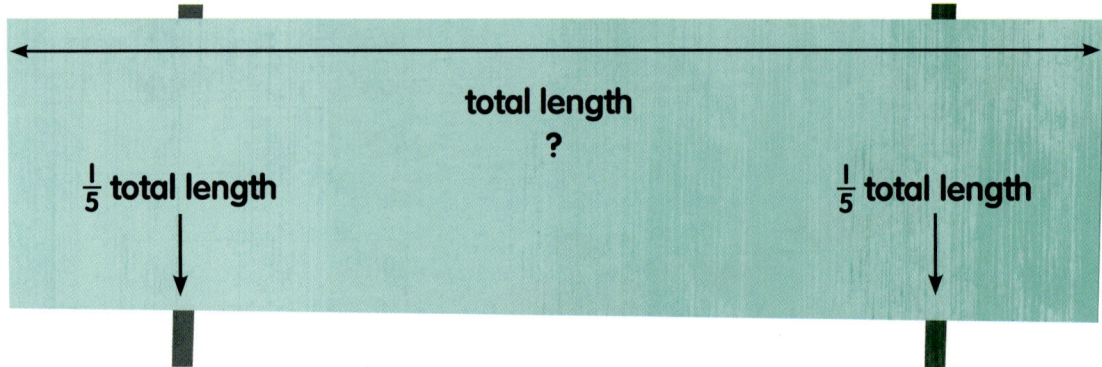

Sophie is making advertising signs for a local football club. Each sign will be held up by two posts, with the middle of each post placed exactly $\frac{1}{5}$ of the way from each end.

Practice

 1 The first sign Sophie makes has a length of 175 cm. Can you find the distance between the middle of the two posts, in centimetres? Can you draw a sketch to explain how you worked this out?

2 The second sign is 2·5 m long. Can you work out the distance between the two posts in metres?

3 The third sign is 1·6 m long. Can you give the distance between the two posts in metres this time?

4 Can you work out the following?

a $\frac{3}{4}$ of 5 kg   b $\frac{1}{3}$ of 4·5 ℓ   c $\frac{5}{6}$ of £9

Going deeper

1 Sophie makes another sign that is 2·7 m long and changes the design so that the posts are placed $\frac{1}{6}$ of the way from the ends. Can you work out the distance between the middle of the two posts?

2 Can you think of a real-life question that could go with each of the following calculations? Give them to your partner to solve.

a $\frac{2}{5}$ of 8 ℓ   b $\frac{3}{8}$ of 6 kg   c $\frac{7}{10}$ of £15

Sharing things equally

We have 7 sandwiches to share equally.

Practice

1 **a** Discuss how the 3 friends can share the sandwiches above equally.

  b Before they start eating, another friend joins them. Can you explain how they can share the sandwiches equally now?

2 How much would each person get if 3 friends shared equally:

  a 10 sandwiches   **b** 20 sandwiches?

Going deeper

1 Next time they meet, the 3 friends get $3\frac{2}{3}$ sandwiches each. How many sandwiches did they share? Can you explain how you worked this out?

2 The 3 friends make a big pile of sandwiches. When they share these equally they have one left over. Before they start eating, one more friend arrives. They share the sandwiches out again, and this time there are no sandwiches left over. Each friend gets more than 5 sandwiches.

Can you work out what is the smallest number of sandwiches they could have made? How did you work this out?

Working with area and perimeter

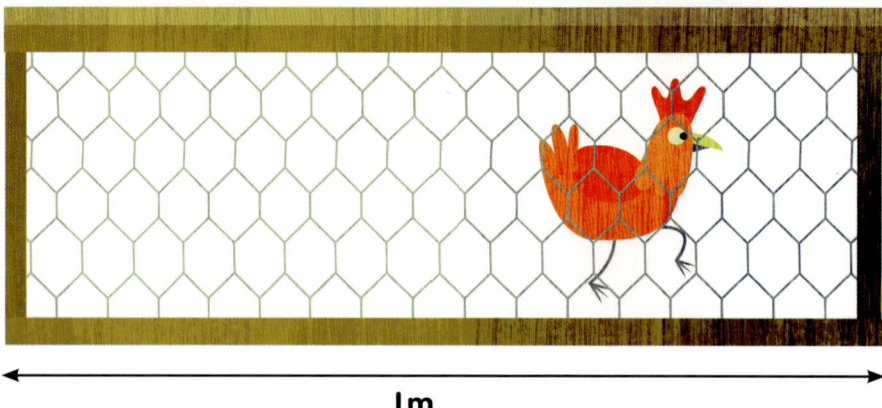

1m

Practice

1 A farmer needs to build a rectangular pen for his 6 chickens, so that each chicken has at least 1m by 1m of space. Fencing panels are 1m long. What is the smallest number of panels he could use?

2 Another farmer has 20m of fencing. How should he arrange it so that his pen can safely contain the largest possible number of chickens? How many chickens could the pen contain?

Going deeper

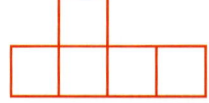

1 A 'pentomino' is a rectilinear shape made from 5 equal squares placed edge to edge. How many different pentominoes can you draw? Which has the largest perimeter?

2 What do you think a 'hexomino' might be?

3 a Can you find the largest possible perimeter of a hexomino made from 1cm squares?

   b Can you explain how you know this is the largest?

Area of composite shapes

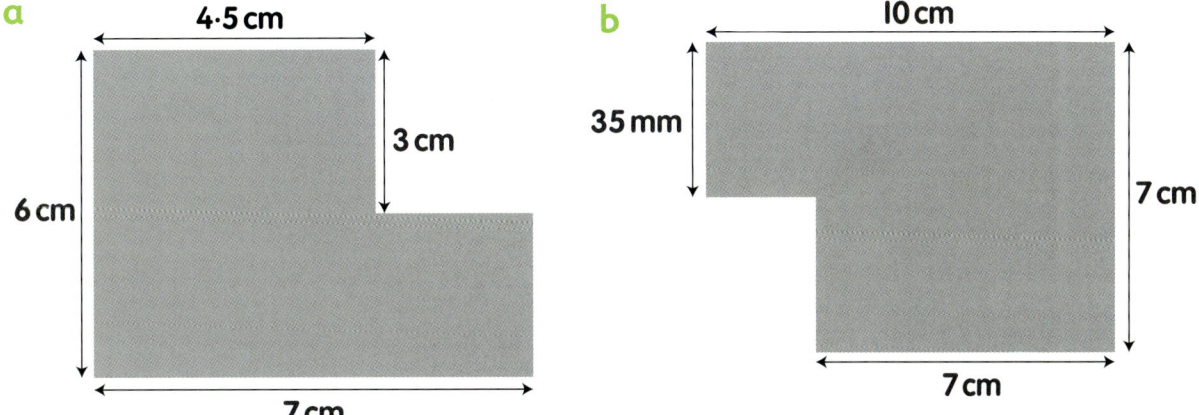

Practice

I Can you find the area of each of these sticky note pads? How did you work them out?

....

Going deeper

I Here are ten 10 cm sticks arranged to make a rectilinear shape.
Can you work out the area of the shape?

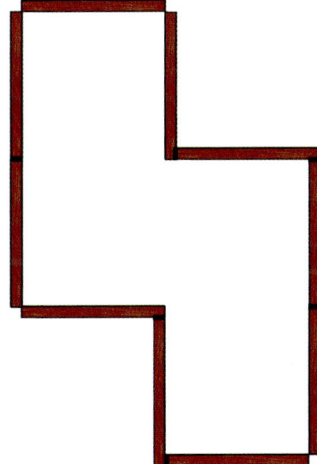

2 a If you were given ten 50 cm sticks, what other rectilinear shapes could you make? Each shape must contain a single area.

b Can you find the area of each shape?

Calculating the area and perimeter of oblongs

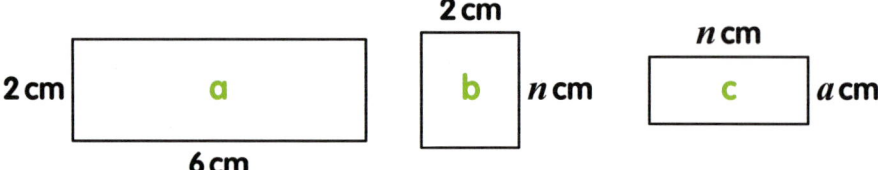

Practice

1 Can you write down the area and perimeter of each of the oblongs above?

2 If a rectangle has an area of 12 cm² and a length of 4 cm, can you work out what its width must be?

Going deeper

1 Can you sketch four rectangles with the following dimensions and label the lengths of their sides? There is no need to find the value of the letters.

a area of $20z$, length of $2z$

b perimeter of $6x + 4$, height of 2

c height of 3, area $6x$

d perimeter of $8 + 4n$, length of $2n$

2 For each answer to **question 1**, can you explain how you worked out the missing information?

Side length, area and perimeter

Practice

1 Look at the enclosed rectangle above, made with two 10-rods and two 4-rods. Can you work out the area and perimeter of the shape in terms of rods?

2 What would the area and perimeter of each rectangle be in terms of rods, if you enclosed a rectangle using:

a two 9-rods and two 5-rods   **b** four 5-rods and two 2-rods

c six 2-rods and two 9-rods?

··

Going deeper

1 Here are some rectangles with the area of each written in the middle. Can you work out the length of each missing side?

a   2

  5   ?

b x

  ? x^2

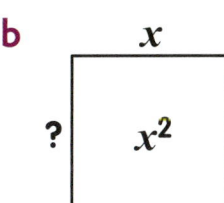

c   5

  $10b$?

d   ?

  12   240

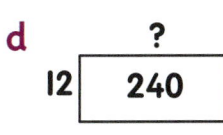

2 a How would you work out the perimeter of each rectangle above?

b For each one, can you find a different rectangle with the same perimeter?

Understanding scale

Practice

1 **a** Mrs Lawson shows a netball team a 1:50 scale drawing of the netball court. What are the dimensions of the real court in metres?

   b Can you work out the area of the netball court?

2 **a** In a picture, Adam appears 6 cm tall. In real life Adam is 1 m 20 cm. What is the scale of the picture?

   b Nima appears in the same picture you can see here, with her height marked. How tall is Nima really?

....

Going deeper

1 A scale drawing shows a railway engine pulling a carriage. The real engine is 5 m long. On the drawing it is shown as 10 cm long. The real carriage is 12 m long. How long will the **drawing** of the carriage be?

2 Angelina is trying to draw an image that is as large and accurate as possible of the Angel of the North statue, in Gateshead, UK. The statue is 20 m high. If her paper is 12 cm in height, what scale factor do you think she should choose? Can you explain your choice?

Making scaled drawings

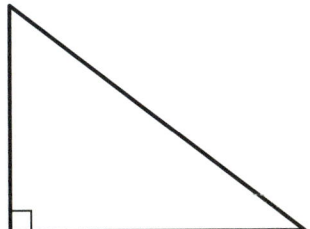

Practice

1 a Can you measure the sides of this triangle and then draw it accurately in your exercise book?

  b Now imagine that it has been scaled **up** by a scale factor of 3 to form a larger triangle. Can you draw the new triangle?

  c How many times can the smaller triangle fit into the larger? Can you explain your reasoning?

2 Choose a room in your school. Use a tape measure to measure the edges of the room, and create a scale drawing using a scale of 1 cm:1 m.

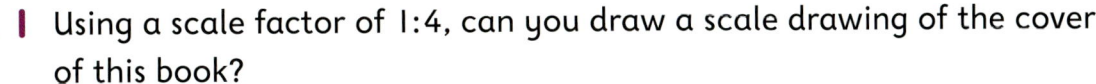

Going deeper

1 Using a scale factor of 1:4, can you draw a scale drawing of the cover of this book?

  If the area of the actual cover is 520 cm², can you estimate the area of your drawing?

  Can you explain your answer?

2 A table-tennis table is 3 m long by 1·5 m wide. Rupesh wants to design an image for the tabletop, so he decides to create a scale drawing of it on a piece of A4 paper.

  What scale factor would you advise him to use?

  Can you explain your reasoning?

Using scale drawings to find actual sizes

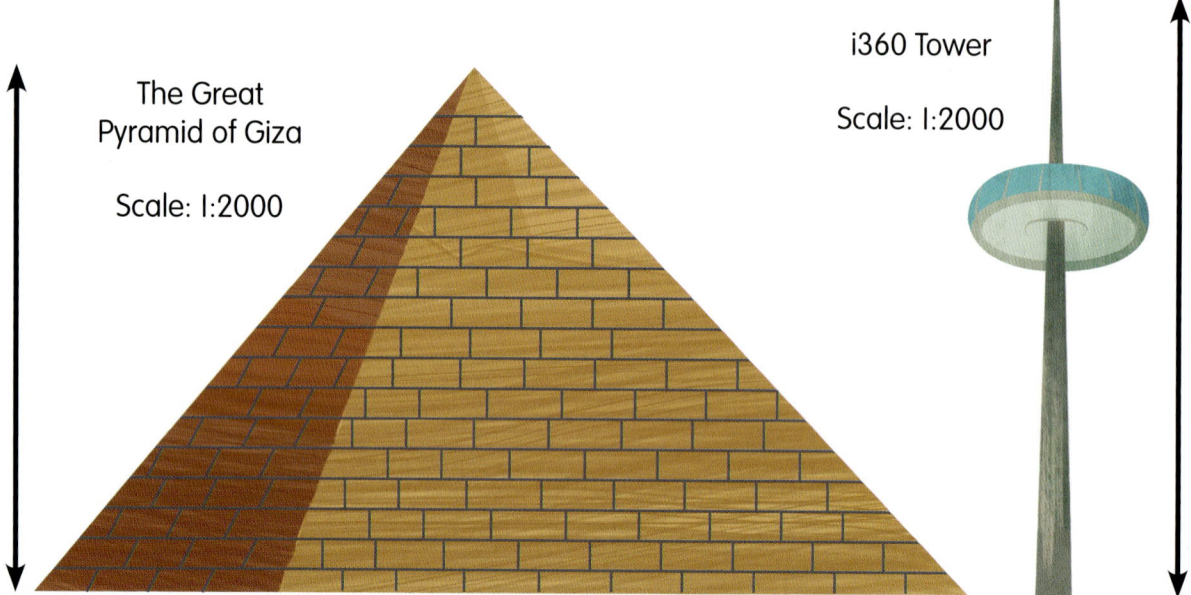

The Great
Pyramid of Giza

Scale: 1:2000

i360 Tower

Scale: 1:2000

Practice

1 Above is a drawing of the Great Pyramid of Giza, Egypt. Can you measure the drawing and use the scale factor to work out the approximate height of the pyramid in metres?

2 The i360 Tower in Brighton, UK is shown in the drawing above. Can you use a ruler and the scale factor to find its approximate height in metres?

Going deeper

1 Lewis lives on a long straight road. The service station is 30 km down the road, from his home. On his map the service station appears 4 cm from his home. What is the scale of the map?

2 Lewis decides not to stop at the services, and keeps on going to his destination, which is 7 cm past the service station on his map. How far will he have travelled altogether when he gets there?

Exploring the effects of scaling

Practice

1 a A roofer is laying panels on a flat, rectangular roof. The roof measures 6 m by 8 m. The panels are each 2 m by 3 m. Using a scale of 1:100, draw a diagram to show how the panels can fit together to cover the roof. How many panels are needed?

b What is the area of each panel?

c What is the area of the roof?

d What is the area of each panel on your diagram?

e What is the area of the roof on your diagram?

2 A gardener has 12 square 1 m by 1 m slabs of turf to create a rectangular lawn in the centre of a garden. Using a scale of 1:50, can you draw scale drawings to show the possible dimensions of the lawn?

Going deeper

1 An oil rig is 24 m high, and on a scale drawing it is shown as 80 cm high. A pump stands next to the oil rig, and this is drawn as 32 cm high. How tall is the pump? Can you explain how you worked it out?

2 a A rectangle with an area of 36 cm^2 is represented on a scale drawing. The area of the diagram is only 4 cm^2. What scale factor was used for the drawing? Can you explain how you worked this out?

b Now choose a different scale factor and draw the rectangle. What has happened to the area of the scale drawing now?

Calculating with fractions

Practice

I a Claire eats $\frac{1}{5}$ of the tray of flapjacks and Tim eats $\frac{2}{5}$. What fraction of the tray of flapjacks has been eaten? What fraction is left?

b Two more friends arrive. Amina eats the same quantity of flapjacks as Tim, and Jay eats $\frac{3}{5}$ of a tray. They have to bake another tray of flapjacks. Can you explain why, using number rods to help you?

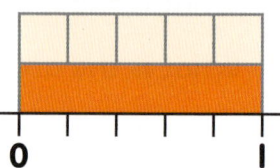

c How many trays of flapjacks have the four children eaten altogether? Can you write this as an improper fraction and as a mixed number?

2 a Can you work out the total of $\frac{5}{8}$ and $\frac{2}{8}$?

b Can you find the difference between $\frac{5}{8}$ and $\frac{2}{8}$?

3 Can you calculate $\frac{7}{8} + \frac{5}{8}$? Can you give the answer as an improper fraction and as a mixed number?

4 a How could you use number rods to show $\frac{12}{9} + \frac{4}{9}$?

b Can you calculate $\frac{12}{9} - \frac{4}{9}$?

Going deeper

 I Can you find two fractions that:

a total $\frac{3}{4}$ **b** have a difference of $\frac{3}{4}$?

2 a Can you find a pair of fractions that total $1\frac{1}{2}$? Now find two more pairs.

b Can you find three pairs of fractions with a difference of $1\frac{1}{2}$?

Adding and subtracting fractions on a number line

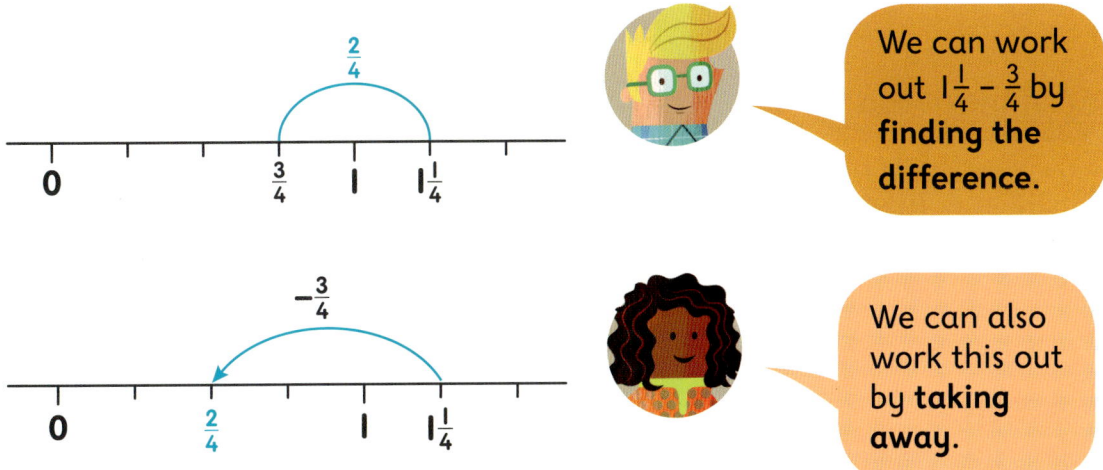

We can work out $1\frac{1}{4} - \frac{3}{4}$ by **finding the difference**.

We can also work this out by **taking away**.

Practice

1 Can you explain how to solve $2\frac{3}{8} - \frac{5}{8}$ by finding the difference? Can you explain how to work it out by taking away?

2 Can you use a number line to show these calculations?

  a $\frac{3}{8} + \frac{7}{8}$ b $\frac{11}{6} - \frac{5}{6}$ c $\frac{12}{5} - \frac{3}{5}$

  For **b** and **c**, can you say whether you solved it by finding the difference or by taking away?

Going deeper

1 a Work with a partner and start with $\frac{3}{8}$. Take turns to add $\frac{4}{8}$ (that is, $\frac{1}{2}$) to the fraction. Can you see a pattern? If so, can you describe it?

  b Now start at 8 and take turns to subtract $\frac{2}{3}$. Describe the pattern.

2 Try this game with a partner:

  • Start at 0.

  • Take turns to add $\frac{1}{4}$, $\frac{2}{4}$ or $\frac{3}{4}$ to the total each time.

  • The winner is the person who reaches the target number of 4.

Further adding and subtracting fractions

I have used $\frac{3}{8}$ of the ribbon.

I have used $\frac{1}{4}$ of the ribbon.

Practice

1 a Can you work out what fraction of the ribbon Ravi and Molly have used altogether?

   b What fraction of the ribbon is left? Can you explain how you know?

2 Can you take each of these fractions away from $\frac{3}{4}$? Use number rods to help you.

   a $\frac{1}{2}$ b $\frac{3}{8}$ c $\frac{5}{12}$

· ·

Going deeper

 1 Choose four pairs of fractions to add together from the group below. Which pairs are easy to add together? Can you explain why?

$$\frac{1}{4} \qquad \frac{2}{3} \qquad \frac{1}{6} \qquad \frac{3}{7} \qquad \frac{2}{5} \qquad \frac{3}{10} \qquad \frac{2}{7} \qquad \frac{5}{8}$$

2 a Ben thinks $\frac{1}{2}$, $\frac{1}{5}$ and $\frac{3}{10}$ together make one whole. Is he correct? Can you explain how you know?

   b Can you find five other sets of three fractions with different denominators that add up to one whole?

Multiplying fractions by whole numbers

Emily's vegetable patch is 5 m long and 5 m wide. She plants a row of lettuces that is $\frac{1}{2}$ m wide.

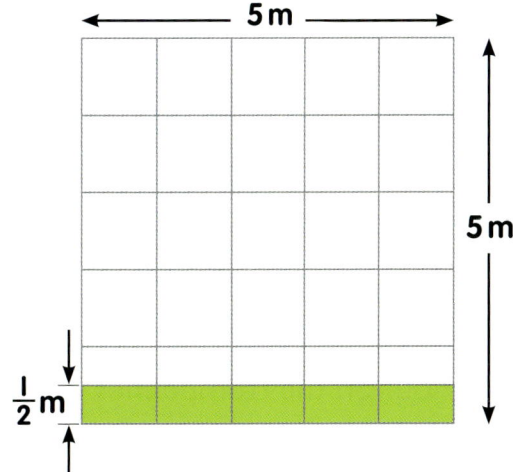

Practice

1 a Can you work out the area of Emily's vegetable patch that is taken up by lettuces?

b Emily also plants carrots in a row $\frac{1}{4}$ m wide. Then she plants peas in a row $\frac{3}{4}$ m wide. Can you work out the area taken up by each type of vegetable?

c Can you work out what area of the vegetable patch is left to plant?

Going deeper

1 Can you work out what multiplying calculation is shown on this number line?

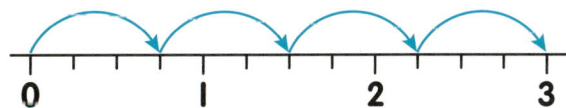

2 Can you draw a number line to show $\frac{1}{3} \times 4$?

3 Can you draw a number line to show some other examples of multiplying fractions? You can ask a partner to work out what each calculation is.

Problem solving

Practice

1 Elsa has £15. If she spends 15 minutes on the go-karts and 45 minutes on the boating lake, how much money will she have left?

2 Nihal has £15. If he spends half an hour on the trampolines and one hour on the climbing wall, how much money will he have left?

3 Can you work out how much it would cost if two children went on each attraction for half an hour? Discuss your strategies with a partner.

Attraction	Price per 15 min
Trampolines	£1·50
Bumper slide	£2·25
Go-karts	£3·15
Climbing wall	£2·80
Boating lake	£3·60

Going deeper

1 a If there are 25 children and they follow the timetable shown, how much would their day at the Fun Park cost?

b If the budget for the trip was £60 per child, how much money would be left over?

Fun Park Timetable:

9 a.m. – 10 a.m.	Climbing wall
10 a.m. – 10:15 a.m.	Bumper slide
10:15 a.m. – 11:00 a.m.	Break
11 a.m. – 12:30 p.m.	Go-karts
12:30 p.m. – 1:30 p.m.	Lunch
1:30 p.m. – 2:45 p.m.	Boating lake
2:45 p.m. – 3 p.m.	Trampolines

2 Elsa spent an hour, in total, doing two thirty minute activities which cost £8·60. Can you work out which activities she did?

Multiplying and adding

Practice

| In the cafe Oli buys an apple juice and an ice cream. How much does he spend?

Apple Juice   55p
Ice Cream   £1·35
Smoothie   70p
Popcorn   £1·10
Water   40p
Fruit Pack   90p
Orange Juice   60p
Cake   £1·15

2 Emily says, "I'm going to buy a smoothie and an ice cream. That will cost £1·95." Is she correct?

3 The class orders the following items. What will the total cost be?

6 apple juices	12 ice creams
8 smoothies	5 popcorns
4 cups of water	3 fruit packs
7 orange juices	5 cakes

Going deeper

| Another class orders the items below at the cafe. When working out the total cost for each item, which calculations can you solve mentally? Discuss when it helps to use written methods.

5 apple juices	11 ice creams
7 smoothies	8 popcorns
6 cups of water	2 fruit packs
4 orange juices	3 cakes

2 Can you write some more calculations for a partner to solve mentally using the price list?

3 If the prices were cut by half, what difference would this make to the overall cost?

Subtracting and dividing

Total cost £5·10

Apple Juice 55p
Ice Cream £1·35
Smoothie 70p
Popcorn £1·10
Water 40p
Fruit Pack 90p
Orange Juice 60p
Cake £1·15

Practice

 1 Three friends each buy an apple juice and the same snack. They spend £5·10 in total. What snack did they buy? Can you explain how you worked this out?

2 Five friends put all their money together and have £10 between them. They want to buy the same snack each and 2 orange juices and 3 smoothies, but they are 5p short.

Can you work out what snack they wanted to buy, and explain your strategy for solving this?

Going deeper

 1 Four children spend £6 exactly. What could they buy so that each child gets the same drink and snack?

List two different solutions.

2 Five children have £8 between them. They all want to buy the same snack and the same drink. What could they buy? Can you list all possibilities and the change they will get with each?

Adding and subtracting

Fun Park attraction	Price per 15 min	Number of staff
Trampolines	£1·50	1
Bumper slide	£2·25	2
Go-karts	£3·15	4
Climbing wall	£2·80	3
Boating lake	£3·60	5

Each member of staff is paid £7·20 per hour and the Fun Park is open for 6 hours per day.

Practice

 1 Can you work out how much profit the bumper slide makes in one day if 50 children spend 15 minutes each on it?

2 During one day, 50 children each spend 15 minutes on each attraction. Which attraction makes the most profit?

 3 Is Tia correct? Work with a partner and discuss your methods.

You may wish to draw a table to help you organize your work.

Some attractions make a loss.

Going deeper

1 Assume that 50 children spend 15 minutes on each attraction every day. Can you change the prices for some attractions so that the park makes between £200 and £250 per day in total? Explain how you worked this out with your partner.

Problem solving and time

My birthday is 18 December.

My birthday is 11 March.

My birthday is 7 February.

My birthday is 12 June.

Ravi
It is 18 March today. My birthday is in 12 weeks and 2 days.

Molly
Ravi's birthday is 3 months and 1 day after mine, so there are 94 days between our birthdays.

Tia
Molly's birthday comes exactly 5 weeks after my birthday.

Ben
312 hours after my birthday, it will be a new year!

Practice

1 a Can you work out if what each child is saying is correct?
      (It is not a leap year.)

   b For anything they said that is not correct, can you say what mistake
      has been made, and how to correct it?

Going deeper

 1 Can you work out:

   a how many seconds there
      are in a day

   b how many hours there
      are in a year

   c how many weeks there are in 67
      days (Hint: write the remainder
      as a fraction.)

   d how old you are in hours?

Converting currencies

Practice

1 a Can you explain what 'currency exchange' means?

b How many Australian dollars would you get in exchange for £1, for £2, for £100 and for £0?

2 Can you draw a conversion graph on graph paper showing pounds and Australian dollars?

3 a Can you say how many Australian dollars you would get for £25? How do you know?

b Can you think of a way to check your answer to **question 3a**?

c How many Australian dollars would you get for £2500? Can you explain how you worked this out?

4 How many pounds would you get for AUS $100? How do you know?

Money Switch
Currency Exchange

For every £1, get:
$1·60 Australian dollars (AUS)
5·20 Malaysian ringgits (RM)

Going deeper

1 At the end of his trip to Malaysia, Anil has RM 480 left to change back into pounds. At the end of her trip to Australia, Becky has AUS $160 left to change back into pounds. Who will get the most after changing their money?

2 Gemma lives in Sydney, Australia and is visiting Kuala Lumpur, Malaysia. She needs to take RM 5600 with her. She has saved AUS $1000. Will this be enough? Can you explain how you know?

Solving money problems

Cola £1·60 — 25% off if you buy 6 or more!

Lemonade £1·70 — Buy 1 get a second half price!

Orangeade £2 — Buy 2 and get a third free!

Practice

1 Which drinks offer is the cheapest if you buy:

   a 2 bottles   **b** 3 bottles   **c** 5 bottles   **d** 9 bottles?

2 **a** Oscar buys 6 bottles of cola, 6 bottles of lemonade and 6 bottles of orangeade. Which 6 bottles are the cheapest?

   b Which 6 bottles have the largest discount?

Going deeper

1 **a** Which taxi company offers better value for a journey of 6 miles?

   b Which company offers better value for a journey of 12 miles?

Calvin's Cabs — 30p per mile

Tanya's Taxis — 50p hire charge then 25p per mile

   c Can you find a distance which both taxis will charge the same amount for?

2 A grocer buys 250 kg of apples for 80p per kilogram. She usually sells them for £2 per kilogram.

   a If she sells all 250 kg of apples at the usual price, how much profit would the grocer make?

   b If she sells them all at half price, how much profit does she make?

Solving volume and capacity problems

strawberry bar

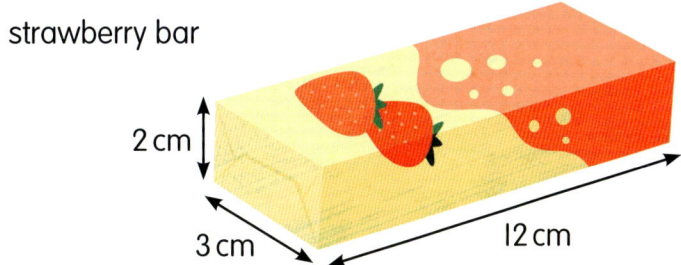

2 cm

3 cm

12 cm

Practice

1 A food company needs to pack strawberry bars into boxes.

What shape is the strawberry bar? Can you find the volume of one bar?

2 The company wants to pack the bars in boxes of 20.

a What capacity does each box need to have?

b The company already has some cuboid-shaped packing boxes with internal dimensions of 5 cm by 12 cm by 24 cm. Can you explain why it cannot use these boxes for the strawberry bars?

c Can you work out dimensions for a cuboid-shaped box that could hold 20 bars exactly (with no empty space left over)? Can you find more than one possibility?

Going deeper

1 a If you wanted to pack exactly 120 identical strawberry bars in a box with a capacity of 7200 cm³, what dimensions might each bar have, in whole centimetres? Can you find all the possibilities?

b Choose one of the sets of dimensions you worked out in **question 1a**. Can you suggest dimensions for the box to pack 120 bars in?

2 Can you design a box with a capacity of exactly 4·5 ℓ? Can you design more than one? (Hint: 1 ℓ = 1000 cm³.)

Using equivalence

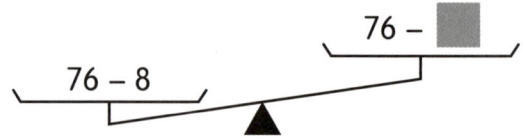

$$76 - \blacksquare$$
$$76 - 8$$

Practice

1 a What is the smallest positive whole number that could go in the box?

b Can you explain how you know, using a number line?

2 Without calculating the amounts below, which do you think is bigger? Can you use inequality signs to write your answers, and explain how you worked these out?

a 234 − 76 or 240 − 83

b 127 + 15 or 112 + 29?

3 Can you explain which is smaller:

a 21 × 6 or 38 × 3

b 84 ÷ 6 or 39 ÷ 3?

4 Can you solve the empty box problems below?

a 347 + 25 > 349 + ▨

b 67 − 18 < 65 − ▨

c 29 × 4 > ▨ × 12

d 32 ÷ 5 < ▨ ÷ 15

Going deeper

1 a Which of the Practice questions did you find easiest?

b Which did you find the hardest? Can you explain why?

2 a What are some possible answers to this problem?

77 − ▨ > 79 − 7

b How could you change the problem to make the possible answers range between numbers 1 and 6 only?

3 Can you make up an empty box problem that has numbers in the range 1 to 7 as the only possible answers?

Balancing number sentences

$$\overbrace{75 + 140} \qquad \overbrace{90 + 125}$$

Practice

1 Can you write the number relationship above as a balancing number sentence? Is it correct? Can you explain how you know?

2 Can you write four different balancing number sentences by adjusting the incorrect one below?

$$275 - 30 = 290 - 45$$

3 By putting numbers in place of the symbols, write as many solutions to the balancing number sentence below as you can.

$$\bigstar + \triangle + \triangle = \blacklozenge \times \odot$$

Going deeper

1 Can you write three other balancing number sentences from the one below? Explain how you worked these out.

$$23 \times 8 = \blacksquare \times \bigcirc$$

2 Choose some pairs of numbers that will work in the balancing number sentence below.

$$\blacktriangle \times 4 = \odot \div 2$$

What can you say about the relationship between the $\blacktriangle$ and $\odot$ pairs of numbers?

3 Can you explain any connections between any dividing balancing number sentence and fractions? Thinking about this balancing number sentence may help you: $2 \div 3 = 4 \div 6$.

Using brackets

Practice

1 Can you write number sentences using brackets, to describe the two number rod illustrations above?

2 Can you use number rods to illustrate the following pairs of calculations?

a $4 + (3 \times 3) = 13$ $(4 + 3) \times 3 = 21$

b $5 \times (3 + 6) = 45$ $(5 \times 3) + 6 = 21$

3 Where do you think the brackets should go in these number sentences?

a $6 - 2 \times 8 = 32$ b $3 \times 4 + 2 \times 5 = 90$

4 Can you solve these problems?

a $(\blacktriangle + \blacktriangle) \times \bullet = 24$ b $(\square + \odot) \times 4 = 48$ c $\blacksquare - (\bigcirc \times 4) = 6$

....

Going deeper

1 Can you make up a number story that would illustrate one of the number sentences in **question 2** above?

2 When Kristina and Max went shopping, three of the things they bought cost £15·50 altogether. If two of these things cost the same, can you write a number sentence using brackets that shows how the items came to this total?

3 Can you explain how to work out this kind of calculation?

$(5 \times 15) - (12 \times 4) = \boxed{}$

4 Can you solve this problem?

$(7 \times \triangle) - (\square \times 4) = 44$

Factor trees

Practice

1 Can you draw factor trees for 45 and 210?

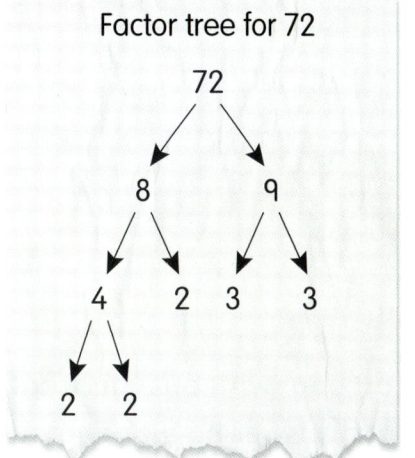

Factor tree for 72

2 **a** Can you write out the prime factors of 45 as a number sentence?

b What do you think is the shortest way of writing this?

3 What do you think is the easiest way to calculate 6 × 3 × 5 in your head? Can you write your answer using brackets?

4 **a** Sally was thinking of a number that had prime factors 2, 2, 3, 3 and 5. What number was she thinking of?

b What do you think is the easiest way to calculate 2 × 2 × 3 × 3 × 5? Can you show your answer using brackets?

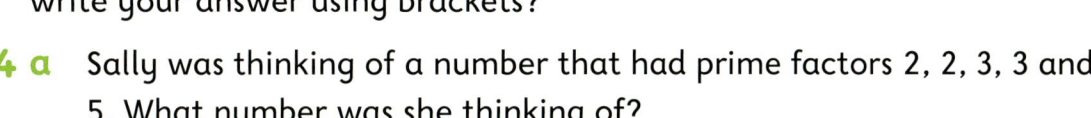

Going deeper

1 Can you work out all the factor pairs of 60? Use brackets and the prime factors of 60 to help you work systematically.

2 What do you think is the easiest way to calculate 16 × 25? Did you use factors? Can you explain?

3 Just using the numbers 2, 3 and 5 as many times as you like, and multiplying them, what numbers can you make up to 150? (For example: 5 × 3 × 2 × 5 = 150.) Can you write these out in order?

Logic and reasoning

Practice

1 Can you work out the total of all the numbers 2 to 8? Can you check your answer by working out the total in another way?

2 Can you solve the adding problem below in two different ways? Can you explain?

$$\frac{1}{5} + \frac{3}{5} + 1 = \blacksquare$$

3 What is your preferred method for solving the problem below? Can you explain why?

$$4·0 + 4·2 + 4·4 + 4·6 + 4·8 = \blacksquare$$

Going deeper

1 How many different ways can you find to calculate these totals?

   a $35 + 42 + 49 = \blacksquare$ b $16 + 24 + 32 = \blacksquare$ c $12 + 18 + 24 + 30 + 36 = \blacksquare$

2 Do you notice anything special about the numbers in each calculation in **question 1**? Can you explain?

3 Can you work out this problem in two different ways? What is your preferred method for solving this?

$$18 + 27 + 36 + 45 = \blacksquare$$

Testing general statements

> If you add two numbers together, you get the same result as when you multiply them.

Practice

1 Do you think Ravi's statement is always, sometimes or never true? Can you explain why, using examples?

2 Can you think of a way of changing Ravi's statement so that it will **never** be true? Can you explain?

3 Do you think Molly's statement is always, sometimes or never true? Can you explain why, using examples?

> The more digits a number has, the larger it is in value.

4 How could you change Molly's statement to make it **always** true? Can you explain?

Going deeper

76	99	77	44
83	51	13	23
24	25	9	16
8	18	27	3

1 Choose any row, column, diagonal or block of four squares and say why one of the numbers in it is the 'odd one out'. Try to do this as many times as you can.

2 "If you add three numbers together you get the same answer as when you multiply them together." Is this statement always, sometimes or never true? Can you explain why, using examples?

3 Can you write a statement about numbers that is **always** true?

4 Can you write a statement about shapes that is **sometimes** true?

5 Can you write a statement about numbers that is **never** true?

Trial and improvement

There are three different Numicon Shapes in the bag and their total is 10. Two of them add up to the third one.

Practice

1 Can you work out which Numicon Shapes are in the bag?
Is this the only solution? How do you know? Can you explain?

2 Can you think of three different Numicon Shapes, where two add up to the third and whose total is 14? How many solutions can you find?

3 For **questions** 1 and 2, did the fact that two Numicon Shapes add up to the third help you to quickly cut down the number of possibilities to explore? Can you explain?

 4 Can you make up another problem with clues like this but choosing numbers from the 0–100 Numeral Cards, for your partner to solve?

Going deeper

1 The numbers missing from these boxes are 1, 3, 5, 6, 7, 8 and 9. Use each number once to make every circle add up to 11. Can you list the numbers in the right order?

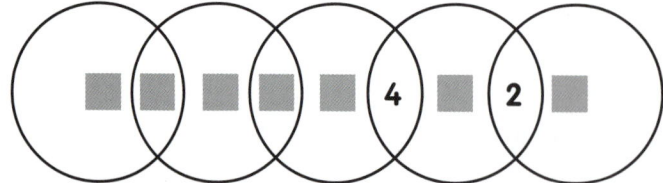

2 The numbers missing from these boxes are 1, 2, 3, 5, 6, 7, 8 and 9. Use each number once to make every circle add up to 13. Can you list the numbers in the right order?

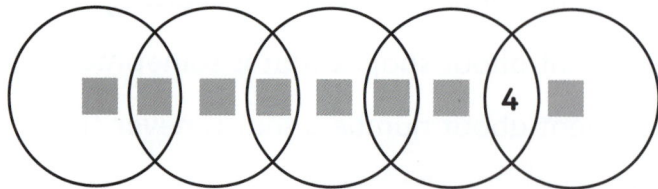

Reasoning about numbers

> I am 10 years old, and in 2 years' time I will be twice as old as my little brother.

HOLLY

Practice

1 How old do you think Holly's brother is now? Can you use number rods to help you explain how you know?

2 Toy cars cost £4. Toy lorries cost £9. Chi spends exactly £48. How many cars and how many lorries does Chi buy?

3 Jeremy's age is now a multiple of 6, and in 3 years' time it will be a multiple of 5. Can you work out how old Jeremy is now?
Can you explain?

Going deeper

1 If the same number goes in each of the empty boxes below, can you work out what number it is? Can you show that you are correct using a double number line?

$$(4 \times \boxed{}) + 2 = (3 \times \boxed{}) + 27$$

2 Archie has three disco lights. The lights are switched on and all three lights come on together. The blue light shines for 2 seconds, then is off for 2 seconds. The red light shines for 3 seconds, then is off for 3 seconds. The green light shines for 5 seconds, then is off for 5 seconds.

a When is the first time that all three lights will be off?

b After the lights are switched on, when is the next time that all three lights will be on together?

Glossary

You can find key words that you need here. Other words that you may have seen before appear in the glossaries of earlier Pupil Books.

2D shape

A flat shape with two dimensions: length and width. (See also **3D shape**, **length**, **width**.)

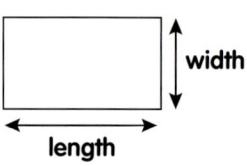

3D shape

A solid shape with three dimensions: length, width and height. (See also **2D shape**, **length, width, height**.)

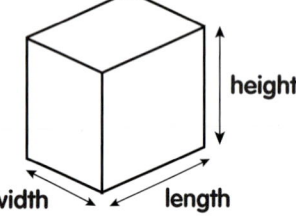

acute angle

An angle smaller than a right angle (90°). (See also **obtuse angle, reflex angle, right angle**.)

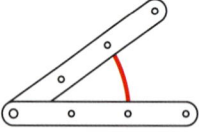

adding

Combining two or more amounts or numbers to make a total. Making something bigger by increasing an amount. (See also **subtracting**.)

angle

An amount of turn or rotation. (See also **acute angle, obtuse angle, reflex angle, right angle**.)

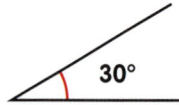

anti-clockwise

The opposite direction to the way the hands on a clock move. (See also **clockwise**.)

area

An amount of surface.

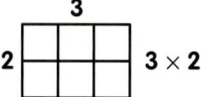

array

A rectangular arrangement of objects or numbers in rows and columns.

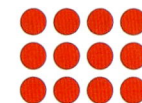

axes

The perpendicular lines which frame a coordinate grid. (See also **perpendicular**, **x-axis, y-axis**.)

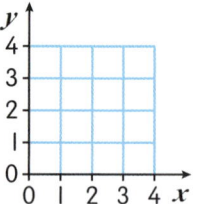

balancing calculation

A calculation that uses the equals symbol to show that two expressions are equal, e.g. 80 + 110 = 140 + 50.

bar chart

A way of representing data using bars.

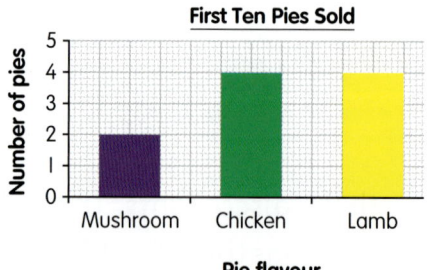

capacity

How much a container can hold measured in, e.g. millilitres (ml) or litres (ℓ). (See also **volume**.)

clockwise

The same direction as the hands on a clock move.
(See also **anti-clockwise**.)

column method

Written method for adding, subtracting, multiplying or dividing in which numbers are written in columns according to their value.

		2	5	6	7
+		2	3	2	6
		4	8	9	3
				1	

common factor

A whole number that divides into two or more other numbers exactly, e.g. 3 is a common factor of 6, 9 and 12.
(See also **factor**, **highest common factor**.)

common multiple

A number that is a multiple of two or more other numbers, e.g. 24 is a common multiple of 2, 3 and 6.
(See also **lowest common multiple**, **multiple**.)

complements

Numbers that add together to make a given total, e.g. 25 and 75, 50 and 50 are complements to 100.

composite number

Any positive whole number that is not a prime number.
(See also **positive number**, **prime number**.)

consecutive numbers

Numbers that follow each other immediately in a sequence.
(See also **sequence**.)

coordinates

Pairs of numbers describing position on a grid.
(See also **axes**, **x-axis**, **y-axis**.)

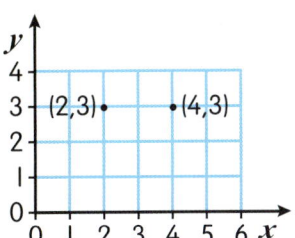

cube number

When a number is multiplied by itself twice, the product is called a cube number, e.g. $2 \times 2 \times 2 = 2^3 = 8$, so 8 is a cube number.
(See also **square number**.)

decrease

A reduction in size or number.

degrees

Unit of measurement used to describe turns, rotations, or temperatures, e.g. 45°.

denominator

Lower number of a fraction, gives the fraction its name.
(See also **numerator**.)

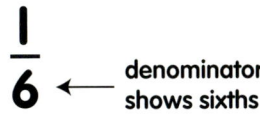

denominator shows sixths

difference

The result of subtracting one number from another, e.g. the difference between 8 and 3 is 5, $8 - 3 = 5$.
(See also **subtracting**.)

digit

A symbol used to represent a number.

dividing

Grouping or sharing a number into equal parts. Or sometimes, 'scaling down', e.g. scaling a recipe for 6 down to a recipe for 2.

double

Multiply a number or amount by two.

edge

Where two faces of a 3D shape meet.
(See also **3D shape**, **face**, **side**.)

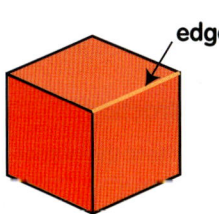

edge

equal

Of the same, or equivalent, value.
(See also **equivalence**, **equivalent**.)

equivalence, equivalent

Different ways of representing the same value, e.g. $6 + 2$ is equivalent to 8, 0.5 is equivalent to $\frac{1}{2}$.
(See also **equivalent fractions**.)

equivalent fractions

Fractions of equal value, represented in different ways, e.g. $\frac{1}{4} = \frac{3}{12}$.
(See also **equivalence**, **equivalent**, **proportion**, **ratio**.)

estimate

Make an informed approximation.

even number

Any whole number that can be divided exactly by two.
(See also **odd number**.)

face

A flat surface on a solid shape.
(See also **3D shape**, **edge**, **side**, **vertex**.)

factor

A number that divides into another number exactly, without leaving a remainder.
(See also **common factor**, **remainder**.)

greater than

Used for comparing values, shown by the symbol >, e.g. 6 > 4 shows that 6 is greater than 4.
(See also **less than**.)

height

The measurement from top to bottom.
(See also **length**, **width**.)

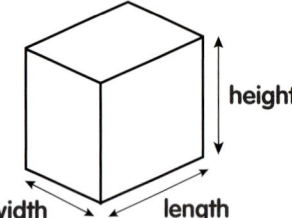

highest common factor (HCF)

The highest number that will divide into two or more other numbers exactly, e.g. 12 is the HCF of 24 and 36.
(See also **factor**, **common factor**.)

improper fraction

A fraction where the numerator is larger than the denominator, e.g. $\frac{9}{6}$.
(See also **denominator**, **numerator**, **proper fraction**.)

integers

Positive and negative whole numbers, including zero.

inverse

The reverse or the opposite. Adding and subtracting have an inverse relation to each other and each can undo the other, e.g. 8 + 6 = 14 so 14 − 6 = 8.

length

The measurement from one end to the other.
(See also **height**, **width**.)

less than

Used for comparing values, shown by the symbol <, e.g. 2 < 4 shows that 2 is less than 4.
(See also **greater than**.)

line graph

A way of presenting data on continuously changing measures, e.g. temperature, water usage.

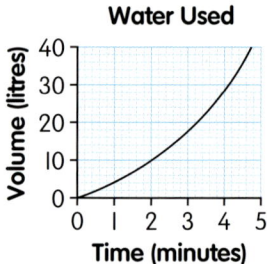

line of symmetry

A line that divides a symmetrical object in half.

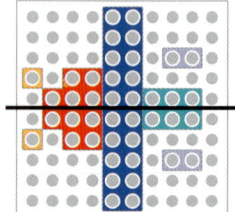

lowest common multiple (LCM)

The lowest number that is a multiple of two or more other numbers, e.g. the LCM of 3, 4 and 6 is 12.
(See also **multiple**.)

mass

The amount of matter in an object measured in, e.g. grams (g), kilograms (kg).

mixed number

A number written as a whole number and a fraction, e.g. $3\frac{2}{5}$.

multiple

The product of two whole numbers larger than one, e.g. 15 is a multiple of 3 and of 5, 5 × 3 = 15.

multiplying

Repeated adding of a number to find 'so many lots of something', e.g. 3 lots of 4 = 4 + 4 + 4 = 3 × 4 = 12. Also 'scaling up', e.g. scaling up a recipe for 2 into a recipe for 6.

negative number

Negative numbers show amounts below zero, e.g. ⁻12.
(See also **positive number**.)

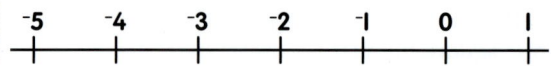

number bond

Shows an adding or subtracting operation between two numbers, along with its outcome, e.g. 8 + 2 = 10.

number rods

Coloured rods of different lengths used for visualizing relationships and calculations.

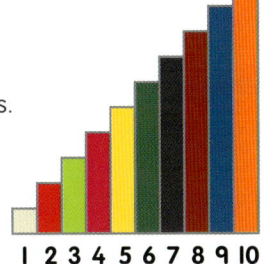

1 2 3 4 5 6 7 8 9 10

numerator

Upper number of a fraction, shows how many of this kind of fraction.
(See also **denominator, proper fraction**.)

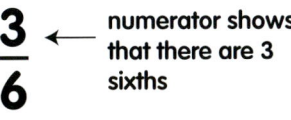

$$\frac{3}{6}$$ ← numerator shows that there are 3 sixths

Numicon Shapes

Shapes of different sizes used for visualizing relationships and calculations.

1 2 3 4 5 6 7 8 9 10

obtuse angle

An angle between 90° and 180°.
(See also **acute angle, reflex angle, right angle**.)

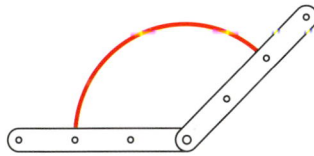

odd number

Any whole number that cannot be divided exactly by two. (See also **even number**.)

partitioning

Splitting a number in different ways, usually to help with calculating, e.g. 27 can be partitioned into 2 tens (20) and 7 units (7).

per cent

Means 'out of 100'. (See also **percentage**.)

percentage

Used to show a fraction 'out of 100' with the symbol %, e.g. 50%.
(See also **per cent.**)

perimeter

The distance around a shape.

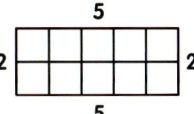

perpendicular

Two lines at right angles to each other.

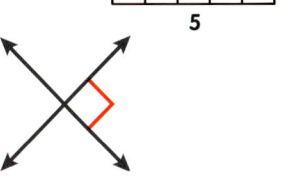

pie chart

A way of presenting data as a 'whole' using fractions of a circle.

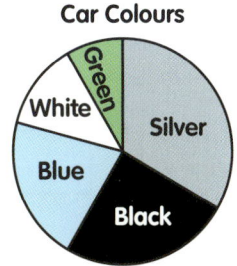

Car Colours

polygon

A flat, geometric shape with straight sides.
(See also **2D shape**.)

positive number

Positive numbers show amounts above zero, e.g. 12. (See also **negative number**.)

prime factors

The smallest parts a composite number can be divided into, e.g. the prime factors of 12 are 2, 2 and 3, because 2 × 2 × 3 = 12.
(See also **composite number, factor, prime number**.)

prime number

A whole number with exactly two different factors, which are 1 and itself, e.g. the only factors of 3 are 1 and 3.
(See also **composite number, factor**.)

product

The number resulting from multiplying two or more numbers together, e.g. in the multiplying calculation 6 × 4 = 24, then 24 is the product.
(See also **multiplying**.)

proper fraction

A fraction where the numerator is smaller than the denominator, e.g. $\frac{1}{10}$. (See also **improper fraction**.)

161

proportion

An expression that shows two ratios or fractions
are equal (in proportion to each other), e.g. $1:2 = 4:8$ or
$\frac{1}{2} = \frac{4}{8}$. Also used to express a fraction of a whole, e.g.
the proportion of grapes in a bag that are green could
be expressed as $\frac{1}{2}$.
(See also **equivalent fractions, ratio**.)

quadrilateral

A polygon with four sides.
(See also **polygon**.)

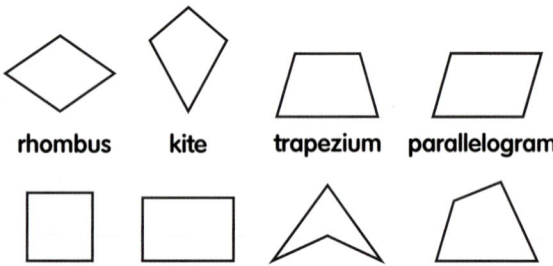

rhombus　　kite　　trapezium　parallelogram

quotient

The number resulting from dividing one number by
another, e.g. in the dividing calculation $24 \div 6 = 4$,
then 4 is the quotient.
(See also **dividing**.)

ratio

A way of comparing two or more quantities
measured in the same units, e.g. if a is 3 times
as much as b this comparison can be written
as the ratio $a:b$ is $3:1$.
(See also **equivalent fractions, proportion**.)

3 : 1

rectilinear shape

A polygon where all the sides
meet at right angles
(See also **polygon**.)

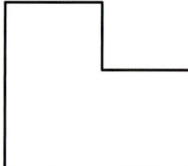

reflection

Transformation of a shape or
point about a line of symmetry
(mirror line).
(See also **symmetry**.)

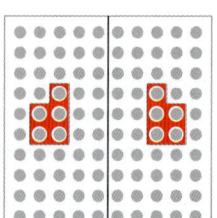

reflex angle

An angle between 180° and 360°.
(See also **acute angle, obtuse
angle, right angle**.)

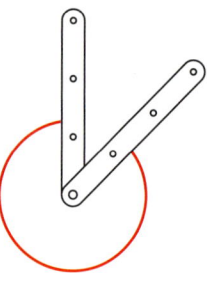

remainder

An amount left over after dividing,
if one number does not divide exactly
by another number, e.g. $110 \div 7 = 15$ r5.
(See also **dividing**.)

right angle

An angle of exactly 90°.
(See also **acute angle, obtuse angle,
reflex angle**.)

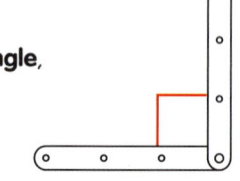

Roman numerals

Symbols used by the Roman to represent numbers,
e.g. $I = 1$, $V = 5$, $X = 10$, $L = 50$, $C = 100$.

rounding (up or down)

Increasing or decreasing a number or amount to
make it closer to (usually) a multiple of ten, or a whole
measuring unit, e.g. rounding or 353 to 350 or 89 cm to
1 metre. Often done to make calculating easier, but less
accurate.
(See also **estimate**.)

scale drawing

An image of a real-life object that has had its
dimensions enlarged or reduced in size using the same
scale factor. A scale drawing is said to be 'in proportion'
to the object it represents.
(See also **proportion, ratio**.)

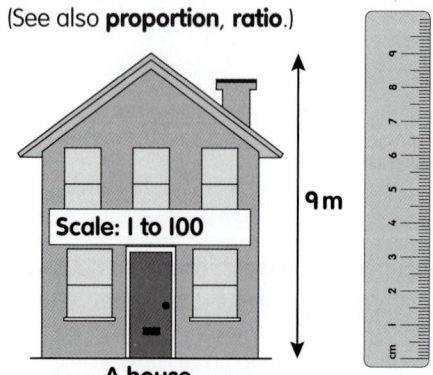

Scale: 1 to 100

9 m

A house

sequence

An ordered list of numbers, shapes or objects (See also **consecutive numbers**.)

side

Straight line joining the vertices of a polygon. (See also **edge, face, polygon, vertex.**)

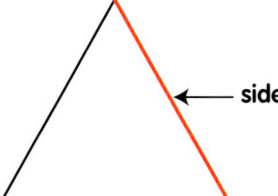

← side

square number

When a number is multiplied by itself, the product is called a square number, e.g. $3 \times 3 = 3^2 = 9$, so 9 is a square number. (See also **cube number**.)

subtracting

Taking one thing away from another, decreasing the size of something, or finding the difference between two numbers. (See also **adding**.)

symmetry

Objects or images with halves that mirror each other are symmetrical, e.g. butterflies, tennis courts. (See also **reflection**.)

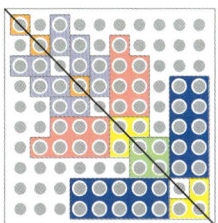

transformation

A way of describing the changes that can be made to the size and/or position of a shape or object, e.g. reflection, translation, rotation or scaling. (See also **symmetry, translation**.)

translation

A transformation involving sliding a shape or object to a different position in a specific direction. (See also **symmetry, transformation**.)

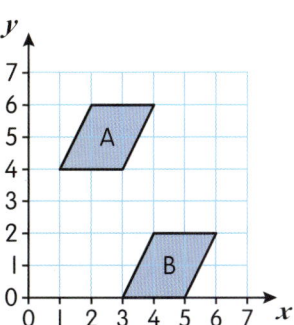

vertex (plural vertices)

A point where two sides meet in a flat shape, or a point where three or more edges meet in a 3D shape. (See also **2D shape, 3D shape, edge, polygon, side**.)

volume

How much space something takes up measured in, e.g. cm^3 or m^3. (See also **capacity, mass, weight**.)

weight

How heavy something is. In everyday life this is often incorrectly described in units of mass; this works because 'weight' is directly proportional to 'mass', i.e. if you double the mass of something you will double its weight. (See also **capacity, mass, volume**.)

width

A measurement from one side to another. When describing oblongs, the width is usually the shorter distance between sides. (See also **length, height**.)

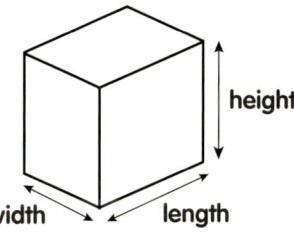

height
width   length

x-axis

The horizontal line which forms part of the frame of a coordinate grid. (See also **axes, perpendicular, y-axis**.)

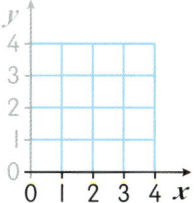

y-axis

The vertical line which forms part of the frame of a coordinate grid. (See also **axes, perpendicular, x-axis**.)

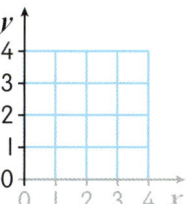